THE TRUTH ABOUT CHERNOBYL

THE
TRUTH ABOUT
CHERNOBYL

Grigori Medvedev

Translated from the Russian
by Evelyn Rossiter

With a Foreword by Andrei Sakharov
and
Author's Preface to the American Edition

BasicBooks
A Division of HarperCollinsPublishers

La Vérité sur Tchernobyl is published with the permission of the Éditions Albin Michel

Library of Congress Cataloging-in-Publication Data

Medvedev. Grigorii.
    [Chernobyl 'skaia khronika. English]
    The truth about Chernobyl / Grigori Mevedev ; translated
from the Russian by Evelyn Rossiter ; with a foreword by Andrei Sakharov and author's preface to the American edition.
        p.   cm.
    Translation of: Chernobyl 'skaia khronika.
    Includes index.
    ISBN 0-465-08775-2
    1. Chernobyl Nuclear Accident, Chernobyl', Ukraine,
1986. I. Sakharov, Andrei, 1921- . II. Title.
TK1362.S65M4313   1991
363.17'99'0947714—dc20   90-55599
                                                     CIP

Édition originale russe:
Чернобцльская хроника
Copyright © 1989 by VAAP
Moscou, 1989
Traduction française:
Copyright © 1990 by Éditions Albin Michel S.A.
22, rue Huyghens, 75014 Paris
ISBN 2-226-04031-5

English translation:
Copyright © 1991 by Basic Books, Inc.
Designed by Ellen Levine
PRINTED IN THE UNITED STATES OF AMERICA
91 92 93 94 CC/HC 10 9 8 7 6 5 4 3

Contents

Foreword

THE TRUTH ABOUT *Chernobyl* is a clear and impartial account of
a tragedy that, three years after the event, still inspires fear and
anxiety. In it, Grigori Medvedev, a nuclear power expert, has provided us with the first complete and objective account of what
happened, devoid of the evasiveness or the omissions found in official versions. Having worked at Chernobyl, he is familiar with the
plant and its staff. In the course of his duties, he attended many
high-level meetings about the construction of nuclear power plants.
Shortly after the disaster, Medvedev was sent to Chernobyl, where
he was able to see the aftermath of the accident and learn a great deal
about what had happened. He gives many technical details necessary for an understanding of these events, reveals the secret work-
ings of the bureaucracy, and denounces the miscalculations of designers and scientists, and pressures from high Soviet officials, as
well as the harm caused, before and shortly after the accident, by the
lack of open communication.

A day-by-day account of the tragic developments of April and
May 1986 at Chernobyl takes up most of this story. The author
describes the behavior and the role of the various characters in this

drama, men of flesh and blood, with their good and their bad points, their doubts, weaknesses, mistakes and heroism, as they reacted to the rampaging nuclear monster. The reader cannot fail to be moved by this book. The remarkable feats of the firefighters have already been widely publicized; now *The Truth about Chernobyl* reveals the heroism of the electricians, the turbine engineers, the operators, and other members of the plant's staff who fought to contain the accident.

Glasnost must apply to every aspect of the Chernobyl disaster, its causes and consequences. The absolute truth must be known. Everyone must be able to form an opinion about a matter that vitally affects our lives and our health, as well as the health of our offspring. Everyone must be entitled to take part in the adoption of decisions that will determine the future of our country and of the world.

Should nuclear power be developed? If so, should the construction of nuclear power stations—even if they are much more reliable than the one at Chernobyl—be permitted aboveground? Or should they be built underground? These issues are so crucial that they cannot be left to technical experts, and still less to bureaucrats, whose approach is too narrowly technical, too tendentious and sometimes prejudiced, as it is paralyzed by a network of mutual solidarity. And the same is true of many other economic, sociological, and ecological issues.

I personally believe that mankind needs nuclear energy. It must be developed, but with absolute guarantees of safety, which means that the reactors must be built underground. International legislation requiring that reactors be built underground must be prepared without delay.

Andrei Sakharov
May 1989

Author's Preface
to the American Edition

THE TRUTH ABOUT *Chernobyl* is being published in the United
States at a time when the antinuclear movement is more vigorous
than ever before throughout the whole world, and particularly in
the Soviet Union.

It now seems hardly surprising that Chernobyl marked the final,
spectacular collapse of a declining era. As a result of Chernobyl, the
nuclear bureaucrats succeeded in creating a new, more covert, and
insidious version of the violence that has been perpetrated in the
Soviet Union for more than seventy years: this new violence—
radiation—was in turn aggravated by a deliberate policy of down-
playing its dangers, as well as by the secrecy that surrounded the
Chernobyl tragedy.

As a witness to these events, and drawing on my experience as
both nuclear power expert and writer, I started work on *The Truth
about Chernobyl* immediately after the explosion, completing my
story in May 1987. In it I sought to answer questions that, I am
convinced, have proved deeply disturbing to many people:

1. By what route did we all arrive at Chernobyl?
2. What actually happened, especially on the night of the explosion and over the next few days?
3. What caused the disaster?
4. What were its consequences?
5. What were the lessons of Chernobyl?
6. What does the future hold?

Yet my main purpose in writing this book was not merely to answer these questions, however important they may be. It was to reconstruct in authentic detail, in the light of the answers to those questions, a picture of the full horror of the Chernobyl nuclear disaster, to revive the dead and the maimed, and have them return to the control room to relive those tragic hours. Millions of readers will, in imagination, share with them the dread they felt as calamity struck and the torments resulting from it—and feel, at the same time, a strengthening of their own desire to go on living, and to survive.

I do not doubt that the anguish felt by the readers of this book, whatever country they live in, will be commensurate with the enormity of the Chernobyl disaster and the severity of its consequences. Readers in the United States, where nuclear power is highly developed, will find *The Truth about Chernobyl* particularly relevant to their present-day concerns.

Although many people were slow to realize it, Chernobyl demonstrated the ignominious failure and the sheer insanity of the administrative-command system. By now they have realized it. It was not until four and a half years after the Chernobyl explosion that Byelorussia was declared an ecological disaster zone. Now even the resolution of the twenty-eighth congress of the Soviet Communist Party, published on 14 July 1990, has told the truth about the disaster; about the arrogance and irresponsibility of a number of leading scientists and of the senior officials of ministries and government departments involved in the development, construction, and operation of nuclear power stations; about the prolonged and unwarranted secrecy surrounding the Chernobyl tragedy; and about the inability of those departments to protect the lives and health of the people.

Each reader of *The Truth about Chernobyl* must understand that Chernobyl was a universal tragedy, that the harm done by it is still going on today, that millions of people are still living on land contaminated with radiation, and that they need help and compassion.

Grigori Medvedev
8 July 1990

1

BEFORE CHERNOBYL

THE MYTH OF SAFETY

"THE LOSS OF the *Challenger* crew and the accident at the Chernobyl nuclear power station have heightened our sense of alarm and been a cruel reminder that mankind is still trying to come to grips with the fantastic, powerful forces which it has itself brought into being, and is still only learning to use them for the sake of progress," said Mikhail Sergeyevich Gorbachev in his statement on Soviet Central TV on 18 August 1986.

This exceedingly sober assessment of the peaceful uses of nuclear energy was the first of its kind in the thirty-five years of the development of nuclear power in the Soviet Union. The Soviet leader's words were unquestionably a sign of the times, and of the wind of purifying truth and change which has swept so powerfully over our country.

Even so, in order to learn the lessons of the past, we must remember that throughout an entire three and a half decades, in the press and on radio and television, our scientists have repeatedly told the general public the exact opposite. The ordinary citizen was made to believe that the peaceful atom was virtually a panacea and the ultimate in genuine safety, ecological cleanliness, and reliability. The whole subject of the safety of nuclear power stations generated much ecstatic enthusiasm.

In 1980, in the monthly magazine *Ogonyok*, Academician M. A. Styrikovich exclaimed, "Nuclear power stations are like stars that shine all day long! We shall sow them all over the land. They are perfectly safe!" And that is precisely what they did.

N. M. Sinev, deputy head of the State Committee on the Utilization of Nuclear Energy, used a homespun image to explain matters to the general reader: "Nuclear reactors are regular furnaces, and the operators who run them are stokers." He thus neatly equated nuclear reactors with ordinary steam boilers and nuclear station operators with stokers who shovel coal into a furnace.

This was in every respect a convenient position to take. First of all, it reassured the public; and secondly, it made it possible to pay the staff of nuclear power stations the same wages as workers at thermal power stations, and in a number of cases even less. Nuclear energy was cheap and straightforward, so the pay could be lower. By the early 1980s the wages at thermal power stations had exceeded those paid to operators at nuclear power stations.

But let us continue to scrutinize the cheerfully optimistic assertions about the complete safety of nuclear power stations.

"The waste products of nuclear power, which are potentially extremely dangerous, are so compact that they can be stored in places isolated from the environment," according to the director of the Physico-Energy Institute, O. D. Kazachkovsky, writing in *Pravda* on 25 July 1984. In actual fact, at the time of the Chernobyl explosion, it turned out that there was nowhere to deposit the spent nuclear fuel. In the previous few decades, no storage facilities for spent nuclear fuel had been built; and one had to be built next to the damaged reactor, in a highly radioactive environment, thus exposing construction and installation crews to severe doses of radiation.

"We live in a nuclear age. Nuclear power stations have proved convenient and reliable in operation. Nuclear reactors are preparing to take on the task of heating cities and built-up areas," O. D.

Kazachkovsky wrote in that same issue of *Pravda*, while forgetting to say that nuclear thermal power stations would be erected near major cities.

One month later, Academician Aleksandr Yefimovich Sheidlin declared in *Literaturnaya Gazeta*, "We were delighted to hear of a remarkable achievement—the start-up of No. 4 reactor, generating one million kilowatts of electricity, at the V. I. Lenin nuclear power station, Chernobyl."

One wonders whether the academician's heart missed a beat when he wrote those words, as it was precisely No. 4 reactor that was destined to explode so resonantly—truly a bolt from the blue sky of guaranteed safety at nuclear power stations.

On another occasion, when a correspondent remarked that the expanded construction of nuclear power stations might alarm the general public, the academician replied, "People can be very emotional about these things. The nuclear power stations in our country are perfectly safe for the populations of surrounding areas. There is quite simply nothing to worry about."

The chairman of the Soviet Union State Committee on the Use of Nuclear Energy,* A. M. Petrosyants, played a major role in advertising the safety of nuclear power stations. In his *From Scientific Search to Nuclear Industry,*† written fourteen years before the Chernobyl explosion, he wrote:

> It must be acknowledged that nuclear energy has a brilliant future. Nuclear energy has definite advantages over conventional energy. Nuclear power stations are entirely independent of sources of raw materials (uranium mines), because nuclear fuel is very compact and can be kept in use for a very long time. Nuclear power stations hold great promise for the use of powerful reactors.

The reassuring conclusion he then reached was that nuclear power stations are clean sources of energy which do not add to environmental pollution.

---

*The State Committee on the Utilization of Nuclear Power is subordinate to the Ministry of Medium Machine Building (see page 38).

†A. M. Petrosyants, *Ot naychnovo poiska k yadernoy promyshlyennosti.* (Moscow, Atomizdat, 1972), p. 73.

Turning then to the question of the extent to which nuclear power was to be developed, and its status beyond the year 2000, Petrosyants focused mainly on the adequacy of stockpiles of uranium ore, leaving completely aside the question of the safety of an extensive network of nuclear power stations located in the most densely populated regions of the European part of the Soviet Union. "The main issue in nuclear energy is how to make the most rational use of the miraculous property of nuclear fuel," he emphasized in that same book. And he was concerned primarily not with safety at nuclear power stations but with the rational use of nuclear fuel. He went on to say:

> The continuing skepticism and distrust felt toward nuclear power stations are caused by exaggerated fear of the radiation danger to the personnel working at the station and, in particular, to the surrounding population.
>
> The operation of nuclear power stations in the Soviet Union and abroad, including the United States, Britain, France, Canada, Italy, Japan, the German Democratic Republic and the Federal Republic of Germany shows that they work with complete safety, provided that the established regimes and the necessary rules are observed. Moreover, one could argue about whether nuclear or coal-fired power stations are more harmful for human beings and the environment.

At this point for some reason, Petrosyants failed to note that thermal power stations can run not only on coal and oil (and that in any case such pollution is local in character and far from lethal) but also on gaseous fuel, vast quantities of which are extracted in the Soviet Union and, as is well-known, transported to a number of destinations, including Western Europe. The conversion of the thermal power stations of the European part of the country to gaseous fuel could completely eliminate the problem of pollution from ash and sulfuric anhydride. Yet Petrosyants managed to turn that problem upside down, too, by devoting a whole chapter of his book to the question of environmental pollution from coal-fired thermal power stations, while remaining silent about actual in-

4

stances of environmental pollution from nuclear power stations, with which he must have been familiar. This was no accidental omission. It was intended to steer the reader toward an optimistic conclusion: "The above-mentioned data on the favorable radiation situation within the vicinity of the Novo-Voronezh and Byeloyarsk nuclear power stations are typical for all the nuclear power stations in the Soviet Union. That same favorable radiation situation is also characteristic of the nuclear power stations of other countries," he concluded, in a display of corporate solidarity with the nuclear industry in foreign countries.

At the same time, Petrosyants must have known that ever since it was first brought on-line in 1964, the first single-loop reactor at the Byeloyarsk nuclear power station had repeatedly broken down: the uranium fuel assemblies had behaved in a most capricious manner; and while repairing them, the staff were exposed to heavy doses of radiation. Radioactive exposure of this sort occurred, virtually without interruption, for almost fifteen years. In 1977, in the second reactor at that same station, also with a single-loop design, 50 percent of the nuclear reactor's fuel assemblies melted down. Repairs lasted one year. The personnel of the Byeloyarsk nuclear power station were quickly exposed to severe doses of radiation, and people had to be brought in from other nuclear power stations to perform hazardous repair work. He also must have been aware that in the town of Melekess, in the Ulyanovsk region, highly radioactive waste was being dumped in fissures far underground; that the British nuclear reactors at Windscale, Winfrith, and Dounreay had been discharging radioactive water into the Irish Sea since the 1950s (and they still are). The list of such facts could be continued. I shall merely note that, at the Moscow press conference on the Chernobyl tragedy on 6 May 1986, Petrosyants shocked a great many people with the following remark: "Science requires victims." That is something one cannot forget.

Now, some more pronouncements on the development of nuclear energy.

Obstacles could naturally be expected during the development of such a new industrial sector. In his *I. V. Kurchatov and Nuclear*

Power, * Y. V. Sivintsev, who like Kurchatov was also an advocate of nuclear power, included some interesting reminiscences about earlier efforts to win public acceptance for the notion of the "peaceful atom," and about the difficulties encountered:

> Opponents of nuclear power abroad and in this country sometimes score "victories" in their fight against innovation. The most widely publicized of these was the cancelation of the nuclear power station in Austria, which was decided on shortly after a strident antinuclear campaign. Western journalists were quick to christen that plant "a billion-dollar mausoleum."† The development of nuclear power in the Soviet Union has also had to surmount certain difficulties. In the late 1950s, the advocates of traditional forms of energy prepared the decision of the Communist Party Central Committee and the Council of Ministers of the Soviet Union for halting construction of the Novo-Voronezh nuclear power station and building in its place a conventional thermal power station; moreover, they came very close to having that decision implemented. The main justification was that nuclear power stations were uneconomical at the time. On hearing of this, Kurchatov dropped everything, went to the Kremlin, and managed to arrange for a new meeting of the senior officials; after a heated discussion with the skeptics, he persuaded the authorities to uphold their previous decisions about the construction of nuclear power stations. One of the secretaries of the CPSU Central Committee then asked him, "What are we going to get?" Kurchatov replied, "Nothing! For some thirty years it's going to be an expensive experiment." Nonetheless, he got his way. It is easy to understand why some of us used terms like *nuclear reactor, human tank,* and even *bomb* to describe him.

The optimistic forecasts and assurances on the part of scientists were, of course, never shared by the operators of nuclear plants, who had to deal with the peaceful atom directly, every day, at their place of work and not in the comfort of some quiet office or laboratory. Throughout all those years, information about breakdowns and mishaps at nuclear power stations was rigorously sifted by the ex-

---

*Y. Sivintsev, *I. V. Kurchatov i yadernaya energetika* (Moscow: Atomizdat, 1980), p. 25.
†Here Sivintsev left out one detail: the Austrian public paid the cost of the nuclear power station through voluntary contributions to the government treasury; afterward, the government, having paid off the contractors, proceeded to mothball the station.

tremely cautious ministries, which divulged only what senior policymakers deemed it necessary to publish. I well recall a landmark event of those days—the Three Mile Island accident, on 28 March 1979, which struck the first serious blow against nuclear power and, in the minds of many but not all, dispelled illusions about the safety of nuclear power stations.

At the time I was a section chief in Soyuzatomenergo, the department of the Soviet Ministry of Energy and Electrification which operates nuclear power stations, and I remember how I and my colleagues reacted to that distressing event. Having worked for many years before in the assembly, maintenance, and operation of nuclear power stations, and realizing from firsthand experience that safety at such plants was literally on the razor's edge or a hair's breadth away from breakdown or disaster, we said at the time, "It had to happen sooner or later. Something like that could happen here, too."

But neither I nor people who had worked previously at nuclear power stations were fully informed about the accident. A detailed account of events in Pennsylvania was provided in the *Information Sheet*, which was distributed to the heads of major government departments and their deputies. The question is: Why was there such a need for secrecy in connection with an accident the whole world knew about? After all, knowledge of negative experiences, if promptly transmitted, can help guarantee that the same mistakes are not repeated. In those days, however, the custom was to keep negative information exclusively for the most senior leaders, while censored versions were passed on to those lower down. Yet even those censored versions gave rise to depressing thoughts about the insidious nature of radiation, should it, despite all precautions, actually escape, and also about the need to make the general public aware of those problems. But in those days it was simply impossible to organize that kind of educational effort, as it would have clashed with the official line about the complete safety of nuclear power stations.

Then, deciding to go it alone, I wrote four short stories about the way people live and work at nuclear power stations. Their titles were: "The Operators," "The Expert Opinion," "The Reactor," and "A Nuclear Tan." When I tried to get them into print, however, the publishers replied, "What do you mean? Academicians are always writing that Soviet nuclear power stations are perfectly safe. Academician Kirillin even intends to start a garden right next to a nuclear power station—and you're writing all this stuff! In the West this kind of thing could happen, but not here!" The senior editor of a leading monthly journal, after praising one of the short stories, even said, "If this had happened in the West, then we would have published it."

Even so, I did manage to get one of the short stories—"The Operators"—published in that journal in 1981. I am glad that I succeeded in warning at least those who managed to read it.

Those were, however, what we call the "stagnant years,"* when things moved at their own sluggish pace, so I shall not anticipate events. After all, everything destined to happen did actually happen. Scientific circles remained calm and unruffled. Sober reminders of the possible environmental hazards of nuclear power stations were treated as an attack on the authority of science.

In 1974, at the annual general meeting of the Soviet Academy of Sciences, Academician A. P. Aleksandrov said in particular, "Our critics claim that nuclear power is dangerous and poses the threat of radioactive contamination of the environment. But what about a nuclear war, comrades? What kind of contamination would occur then?"

The logic behind this remark is truly astounding.

Ten years later, at a meeting of the active party members in the Soviet Ministry of Energy, one year before Chernobyl, the same A. P. Aleksandrov commented, on a melancholy note, "Fate has been kind to us, comrades, in that we haven't had a Pennsylvania of our own. Yes, I mean it."

This shows a striking evolution in the thinking of the president of the Soviet Academy of Sciences. Ten years, of course, is a long time, and Aleksandrov can certainly be credited with anticipating the onset of disaster. A great deal had been going on in the nuclear industry: some grave irregularities and breakdowns had occurred; there had been an unprecedented rise in generating capacity; pressures associated with prestige projects had also increased; and the sense of responsibility of those working in the nuclear power indus-

*That is, the Brezhnev period (1966–82).—Ed.

try was clearly beginning to slacken. One wonders, of course, how they could be expected to show a keen sense of responsibility, when at the nuclear power stations themselves everything was apparently so simple and safe.

It was approximately during those same years that changes began to occur in the personnel running the nuclear power stations, as the shortage of operators suddenly grew acute. Those seeking such employment had previously been, for the most part, real enthusiasts, with a profound passion for nuclear energy; whereas now all kinds of people were pouring in. Of course, they were attracted primarily not by the pay, which was not particularly good, but by the lure of prestige. There were people who had earned good money in some other field but had not yet held posts in the nuclear industry. After all, for many years everyone had been saying it was safe! So what are we waiting for? Out of the way, you nuclear experts! Step aside and make room at the important nuclear pie for brothers-in-law and the well connected! And the nuclear experts were shoved out of the way. But we shall come back to that later—after some details about what happened in Pennsylvania, the precursor of Chernobyl.

THE ACCIDENT
AT THREE MILE ISLAND

On 6 April 1979, the United States publication *Nuclear News* reported that, early on the morning of 28 March 1979, a serious accident had occurred in the No. 2 reactor, with a capacity of 880 megawatts (electric), at the Three Mile Island nuclear power station, situated just over seven miles from the city of Harrisburg, Pennsylvania, and belonging to the Metropolitan Edison Company.

The United States government immediately proceeded to investigate all the circumstances of the accident. On 29 March, senior members of the Nuclear Regulatory Commission were invited to the Subcommittee on Energy and the Environment of the House Interior Committee to take part in the investigation into the causes of the accident and in the elaboration of cleanup procedures and measures to prevent a recurrence. At the same time a thorough checkup was ordered for eight reactors in nuclear power stations at

Oconee, Crystal River, Rancho Seco, Arkansas Nuclear One, and Davis-Besse (in South Carolina, Florida, California, Arkansas, and Ohio, respectively). The equipment in use at these reactors and at Three Mile Island had been manufactured by Babcock & Wilcox. At the time, in April 1979, out of the eight reactors, all with practically identical design features, only five were in operation; the others were undergoing scheduled maintenance.

It turned out that the No. 2 reactor at the Three Mile Island nuclear power station was not equipped with a supplementary safety system, although several of that plant's reactors did have such systems. The Nuclear Regulatory Commission required that the equipment and working regimes of each and every reactor unit manufactured by Babcock & Wilcox should be tested. The official from the commission who was in charge of licensing the construction and operation of nuclear facilities stated at a press conference on 4 April that all necessary safety measures would be applied immediately at every nuclear power station in the country.

The social and political repercussions of the accident were considerable. It caused intense alarm not only in Pennsylvania but in many other states. The governor of California raised the possibility of closing down the Rancho Seco nuclear power station, with a capacity of 913 megawatts (electric), near Sacramento, until all the causes of the Three Mile Island accident were thoroughly clarified, and measures taken to prevent a recurrence of such an event.

The official position of the United States Department of Energy was that the public needed to be reassured. Two days after the accident, Energy Secretary James Schlesinger announced that it was the first time such a thing had happened in all the years commercial nuclear reactors had been in operation, and that events on Three Mile Island had to be viewed objectively, avoiding excessive emotion and hasty conclusions. He stressed that the program for the development of nuclear power would be continued with a view to making the United States independent of foreign sources of energy.

According to Schlesinger, both the extent and the magnitude of radioactive contamination in the area around the nuclear power station were "extremely limited," and there was no need for the population to worry. Meanwhile, on 31 March and 1 April, out of the 200,000 people living within a radius of 21 miles of the plant, about 80,000 had already left their homes. People refused to believe the representatives of the utility, Metropolitan Edison, who had tried to persuade them that nothing terrible had happened. On instructions from Governor Richard Thornburgh, an emergency evacuation plan was drawn up for the entire population of the area. Seven schools were closed down in the vicinity of the plant. The governor ordered the evacuation of all pregnant women and children of preschool age living within a radius of 5 miles, and recommended that everyone living within a radius of 10 miles stay indoors. These measures were taken on the advice of Joseph Hendrie, the chairman of the Nuclear Regulatory Commission, after detection of a leak of radioactive gases into the atmosphere. The most critical phase occurred on 30 and 31 March and on 1 April, when a huge hydrogen bubble formed inside the reactor building and was in danger of both exploding and breaching the containment. If that had happened, the whole of the surrounding area would have been exposed to intense radioactive contamination.

In Harrisburg, the American Nuclear Insurers Company set up an emergency branch, which had paid out $200,000 in compensation by 3 April.

On 1 April, President Jimmy Carter visited Three Mile Island. He appealed to the population to follow all instructions, calmly and precisely, should the need for evacuation arise.

On 5 April, during a speech on energy problems, President Carter dwelt in detail on alternative methods such as solar energy, the processing of shale, or the gasification of coal, while saying nothing at all about nuclear energy, whether fission or fusion.

Many senators stated that the accident could bring about "an agonizing reappraisal" of attitudes to nuclear energy, though they felt that, in the absence of any alternative, the United States would continue to have to produce electricity from nuclear power stations. The ambiguous stance taken by the senators on this matter well illustrates the dilemma confronting the United States government after the accident.

The first signs of trouble were noticed at 4 A.M. when, for unknown reasons, the supply of feedwater from the main pumps to the steam generators was cut off. All three emergency pumps, designed specially to maintain a constant supply of feedwater, had been undergoing maintenance for two weeks—a serious violation of the operating rules for nuclear power stations.

As a result, the steam generator was left without feedwater and was unable to remove heat generated by the reactor from the primary loop. As the parameters for steam had been exceeded, the turbine automatically shut itself off. In the primary loop of the reactor vessel, both water temperature and pressure rose sharply. Through the pilot-operated relief valve on the pressurizer, a mixture of superheated water and steam began to be discharged into a special tank. However, once the water pressure in the primary loop had dropped to the normal level of 2,275 psi (160 kg/cm^2), the valve failed in the open position, causing pressure in the tank also to rise above the normal limit. The emergency membrane on the tank disintegrated, sending about 97,700 gallons (370 m^3) of hot radioactive water onto the floor of the containment building (into the central chamber).

The drainage pumps automatically turned on and started to pump the accumulated water into tanks located in the auxiliary building. The staff should have immediately switched off the drainage pumps, to keep all the radioactive water within the containment, but failed to do so.

The auxiliary building contained three tanks, but all the radioactive water entered only one of them. The tank overflowed, covering the floor to a depth of several inches. The water began to evaporate, and radioactive gases mixed with steam entered the atmosphere through the ventilation stack of the auxiliary building; this was one of the principal causes of subsequent radioactive contamination of the surrounding area.

When the relief valve opened, the emergency reactor core protection system began to function by lowering the control rods, thus halting the chain reaction and practically shutting down the reactor. The process of the fission of uranium nuclei in the fuel rods stopped; but the nuclear decay of fragments continued, releasing heat equivalent to about 10 percent of rated power or approximately 260 megawatts (thermal).

As the relief valve was still open, the pressure of the coolant water in the reactor vessel soon fell, and the water was fast evaporating.

The water level in the reactor vessel was falling, while the temperature rose sharply. This apparently led to the formation of a steam-water mixture which caused the main circulation pumps to fail.

As soon as the pressure had fallen to 159 psi (11.2 kg/cm^2), the emergency core cooling system came into operation, and the fuel rods began to cool. This happened about 2 minutes after the start of the accident.*

For reasons that have still not been explained, 4 minutes and 30 seconds after the start of the accident the operator turned off two pumps controlling the emergency core cooling system. He evidently assumed that the whole of the upper part of the reactor core was under water. It seems likely that the operator incorrectly read the dial for the water pressure inside the primary loop and decided that there was no need for the emergency core cooling system. Meanwhile, however, water continued to evaporate from the reactor. The relief valve had apparently failed in the open position; and the operators, using remote control equipment, were unable to close it. As the valve was located in the upper part of the pressure vessel, beneath the containment, it was practically impossible either to open or to close it manually.

The valve stayed open for such a long time that the water level in the reactor fell, and one third of the core was left without any coolant.

In the opinion of experts, shortly before the emergency core cooling system was switched on, or perhaps soon afterward, at least 20,000 fuel rods out of a total of 36,000 (177 fuel assemblies each containing 208 rods) were left without coolant. The protective zirconium cladding around the fuel rods began to crack and fall apart. Highly radioactive fission products started to leak from the damaged fuel rods. The water in the primary loop became even more radioactive.

When the upper sections of the fuel rods became exposed, the temperature inside the reactor vessel climbed above 752°F (400 °C), and the dials on the control panel went off the scale. The computer monitoring the temperature in the core began to display

---

*At this point the situation resembles Chernobyl 20 seconds before the explosion—but at Chernobyl the staff had earlier turned off the emergency core cooling system.

only question marks and continued to do so for the next 11 hours.

Eleven minutes after the start of the accident, the operator again switched on the emergency core cooling system, which he had previously turned off.

In the next 50 minutes the fall in pressure inside the reactor stopped, but the temperature kept on rising. The pumps supplying water to the emergency core cooling system started to vibrate violently, whereupon the operator, evidently fearing that they would be damaged, switched off all four pumps—two of them 1 hour and 15 minutes after the start of the accident, and the other two 25 minutes later. At 5:30 P.M., the main feedwater pump, which had been switched off at the beginning of the accident, once again began to function. Circulation of water in the core resumed. Water once again covered the upper sections of the fuel rods, which had been left without coolant and had been disintegrating for 11 hours.

On the night of 28–29 March, a bubble of gas began to form in the upper part of the reactor vessel. The core became so hot that, because of the chemical properties of the zirconium cladding around the fuel rods, the water molecules began to split into hydrogen and oxygen. The bubble, measuring about 1,060 cubic feet (30 m³) and consisting mainly of hydrogen and radioactive gases—krypton, argon, xenon, and others—greatly hindered the circulation of coolant water, as pressure inside the reactor was now so high. The main danger, however, was that the mixture of hydrogen and oxygen might explode at any moment. The force of the explosion would have been the equivalent of three tons of TNT, which would definitely have destroyed the reactor vessel. Otherwise, the mixture of hydrogen and oxygen could have leaked upward from the reactor and accumulated beneath the dome of the containment building. If it had exploded there, all the radioactive fission products would have entered the atmosphere. The radiation level within the containment had by then reached 30,000 bers* per hour, or 600 times the lethal dose. And that is not all: if the bubble of gas had continued to grow, it would have gradually extracted from the reactor vessel all the coolant water; then the temperature would have increased to such

---

*Biological equivalent roentgen (ber), which is the same as rem, is a unit of radiation exposure often used in Soviet radiological literature.

an extent that the uranium would have melted, as happened at Chernobyl.

On the night of 29–30 March, the size of the gas bubble was reduced by 20 percent; and by 2 April, it was down to a mere 50 cubic feet (1.4 m³). In order to remove the bubble completely and eliminate the danger of an explosion, the technicians used the method known as degassing of water. The coolant water circulating in the primary loop was injected into the pressurizer (by now the relief valve, had, for unknown reasons, closed). In this way, dissolved hydrogen was extracted from the water. Then the coolant water again entered the reactor, where it absorbed part of the hydrogen from the gas bubble. As the oxygen dissolved in the water, the size of the gas bubble gradually diminished. Outside the containment building, a device known as a recombination unit, to convert hydrogen and oxygen into water, was standing ready, having been specially delivered to the facility.

Once the supply of feedwater to the steam generator was restored and the coolant was circulating again in the primary loop, the normal removal of heat from the core resumed.

As we have seen, very intense radioactivity with long-lived isotopes had been allowed to accumulate beneath the containment, and the continued operation of the facility did not make economic sense. On the basis of preliminary data, the cost of the cleanup was estimated at $40 million.* The reactor was shut down for a long time. A commission was set up to inquire into the causes of the accident.

Representatives of the public accused Metropolitan Edison of rushing to bring No. 2 reactor on-line on 30 December, 25 hours before the New Year, in order to save $40 million in taxes, although shortly before that, at the end of 1978, irregularities had already been noticed in the functioning of mechanical devices, and it had been necessary to stop the reactor several times during the test phase. Even so, the federal inspectors allowed commercial operation of the plant. In January 1979, just after it had been brought on-line, the reactor had been shut down for two weeks, when leaks were detected in pipework and pumps.

Even after the accident, Metropolitan Edison continued to com-

---

*The cleanup at Chernobyl cost 8 billion roubles.

mit serious breaches of the safety rules. For example, on Friday, 30 March, on the third day after the accident, 172,000 cubic feet (52,000 m³) of radioactive water were discharged into the Susquehanna River. The utility did this without first getting permission from the Nuclear Regulatory Commission, allegedly in order to make room for more highly radioactive water extracted from the containment building by the drainage pumps.

PRECURSORS OF THE DISASTER

Now, after studying the details of the Pennsylvania disaster, and in anticipation of Chernobyl, we would do well to look back over the past thirty-five years, and inquire whether Three Mile Island and Chernobyl really were random events; whether, since the early 1950s, accidents occurred at nuclear power stations in the United States and the Soviet Union that could have served as a lesson and jolted everyone out of their relaxed approach to an extremely complex problem—the development of nuclear energy.

Have nuclear power stations really worked smoothly in both countries all that time? Not quite, it seems. A glance at the history of the development of nuclear energy will show that accidents began to occur in nuclear reactors virtually as soon as they started operations.

In the United States

1951. Detroit. Accident in a research reactor. Overheating of fissionable material because permissible temperatures had been exceeded. Air contaminated with radioactive gases.

24 June 1959. Melt of part of fuel rods due to failure of cooling system at experimental power reactor in Santa Susanna, California.

3 January 1961. Steam explosion at an experimental reactor near Idaho Falls, Idaho. Three people killed.

5 October 1966. Partial core melt due to failure of cooling system at the Enrico Fermi reactor, near Detroit.

19 November 1971. Almost 53,000 gallons (200,000 liters) of water contaminated with radioactive substances from an overflowing waste storage tank at Monticello, Minnesota, flowed into the Mississippi River.

28 March 1979. Core melt due to loss of cooling at the Three Mile Island nuclear power station. Radioactive gases released into the atmosphere and liquid radioactive waste discharged into the Susquehanna River. Population evacuated from vicinity of disaster.

7 August 1979. About one hundred people received a radiation dose six times higher than the normal permissible level due to the discharge of highly enriched uranium from a plant producing nuclear fuel near the town of Irving, Tennessee.

25 January 1982. A broken tube in a steam generator in the R. E. Ginna nuclear power plant, near Rochester, New York, caused the release of radioactive steam into the atmosphere.

30 January 1982. Emergency declared at a nuclear power station near the town of Ontario, New York. A breakdown in the cooling system caused a leak of radioactive substances into the atmosphere.

28 February 1985. At the Virgil C. Summer nuclear power station, in Jenkinsville, South Carolina, the reactor became critical too soon, leading to an uncontrolled nuclear power surge.

19 May 1985. At the Indian Point 2 nuclear power station, near New York City, owned by the Consolidated Edison Company, there was a leakage of radioactive water. The accident was caused by the failure of a valve and led to the discharge of several hundred gallons of radioactive water, some of which entered the environment outside the facility.

1986. Webbers Falls, Oklahoma. Explosion of a tank containing radioactive gas at a uranium enrichment plant. One person killed. Eight injured.

In the Soviet Union

September 1957. Accident at reactor near Chelyabinsk. A spontaneous nuclear reaction occurred in spent fuel, causing a substantial release of radioactivity. Radiation spread over a wide area. The contaminated zone was enclosed within a barbed wire fence, and ringed by a drainage channel. The population was evacuated and the topsoil removed; livestock was destroyed and buried in pits.

7 May 1966. Prompt neutron power surge at a nuclear power station with a boiling-water nuclear reactor in the town of Melekess. A dosimetrist and the shift foreman at the station were exposed to radiation. Two bags of boric acid were dropped into the reactor to extinguish it.

1964–79. Over a period of 15 years, the fuel assemblies in the core of No. 1 reactor at the Byeloyarsk nuclear power station were persistently damaged by overheating. During repair work on the core, the operational staff were overradiated.

7 January 1974. Explosion of reinforced concrete gasholder for the retention of radioactive gases in No. 1 reactor of Leningrad nuclear power station. No casualties.

6 February 1974. Rupture of intermediate loop in No. 1 reactor at the Leningrad nuclear power station due to boiling of water, with consequent cavitation.* Three people killed. Highly radioactive water with pulp from filter powder discharged into the environment.

October 1975. Partial destruction of the core (local melt) at No. 1 reactor of the Leningrad nuclear power station. The reactor was halted; and within 24 hours, as an emergency cooling measure, liquid nitrogen was pumped into the core and then discharged into the atmosphere through the ventilation stack. About one and a half million curies of highly radioactive radionuclides were discharged into the environment.

1977. Melt of half of the fuel assemblies in the core of No. 2 reactor, at the Byeloyarsk nuclear power station. Repairs, exposing the staff to radiation, lasted about a year.

31 December 1978. No. 2 unit at the Byeloyarsk nuclear power station was heavily damaged by a fire started when a roof panel in

*For an explanation of cavitation, see pages 67–68.

the turbine hall fell onto a fuel tank. The entire control cable was burned out. The reactor was out of control. In the effort to supply emergency cooling water to the reactor, eight persons were exposed to severe doses of radiation.

September 1982. Destruction of the central fuel assembly of No. 1 reactor at the Chernobyl nuclear power station due to errors by the operational staff. Radioactivity was released into the immediate vicinity of the plant and into the town of Pripyat, and the staff doing repair work were exposed to severe radiation while eliminating this "minor irregularity."

October 1982. Explosion of generator in No. 1 reactor of the Armyanskaya nuclear power station. The turbine hall burned down. The operational staff organized the supply of cooling water to the reactor. An emergency group flown in from the Kolsk nuclear power station helped the operators save the core.

27 June 1985. Accident in No. 1 reactor of the Balakovo nuclear power station. During start-up activities a relief valve burst, and steam at 572°F (300°C) entered a room where people were working. Fourteen people were killed. This accident was due to errors made in haste and nervousness by inexperienced operational staff.

SILENCE AND INCOMPETENCE
IN HIGH PLACES

None of the nuclear power station accidents in the Soviet Union were publicly reported, with the exception of those at the first reactors of the Armyanskaya and Chernobyl facilities in 1982; these were hinted at in the leading article of *Pravda*, after the election of Yuri V. Andropov to the post of general secretary of the Communist Party Central Committee.

Apart from that, an indirect reference was made to the accident in No. 1 reactor at the Leningrad nuclear power station at the March 1976 meeting of the active party members of the Ministry of Energy, which was addressed by the chairman of the Council of Ministers, Alexei N. Kosygin. Among other things, he said that the governments of Sweden and Finland had asked the Soviet government for information about the increase in radioactivity over their countries. Kosygin also said that the Central Committee and the Council of Ministers were drawing the attention of everyone in the energy industry to the particular need to ensure that Soviet nuclear power stations complied with nuclear safety and quality standards.

The concealment from the general public of accidents at nuclear power stations had become a standard mode of behavior under the Minister of Energy and Electrification, P. S. Neporozhny. But accidents were hidden not only from the general public and the government but also from the people who worked at Soviet nuclear power stations. This latter fact posed a special danger, as failure to publicize mishaps always has unexpected consequences: it makes people careless and complacent.

Neporozhny's successor as minister, Anatoly Ivanovich Mayorets, who was quite incompetent in energy matters generally, and in nuclear energy in particular, maintained the tradition of silence. Only six months after his appointment, he signed an order of the Ministry of Energy, dated 19 May 1985, stipulating as follows: "Information about the unfavorable ecological impact of energy-related facilities (the effect of electromagnetic fields, irradiation, contamination of air, water, and soil) on operational personnel, the population, and the environment shall not be reported openly in the press or broadcast on radio or television." In his first few months as minister, Mayorets made this morally dubious policy the foundation of his activities.

It was in this carefully contrived "accident-free" atmosphere that Petrosyants wrote his many books and, without fear of being exposed, advertised the absolute safety of nuclear power stations. Mayorets was operating within the framework of a long-standing system. Having first protected himself by means of his notorious "order," he then set about managing the nuclear power industry.

But a department like the Soviet Ministry of Energy, which permeates virtually the whole Soviet economy, has to be managed competently, wisely, and carefully—in other words, morally, bearing in mind the potential dangers of nuclear energy. As Socrates once said, "A man is wise in what he knows well."

How could a man who was totally unfamiliar with this complex and hazardous industry possibly manage nuclear energy? We have a Russian saying—"It doesn't take gods to bake earthenware pots" (meaning that one doesn't have to be a genius to perform certain tasks)—which various high officials were fond of quoting when challenged on the issue of competence in nuclear matters. Here, however, we are dealing not with earthenware pots but with nuclear reactors, which can themselves become very hot indeed.

Nonetheless, Mayorets busily set about dealing with matters wholly beyond his understanding and—following the example set by the deputy chairman of the Council of Ministers, Boris Yevdokimovich Shcherbina, who had promoted him to his post—started "baking nuclear pots." Mayorets's first act on taking up his ministerial appointment was to abolish the central directorate within the Ministry of Energy which handled design and research, and thereby stifled that important engineering and scientific sector. Then, by cutting back on maintenance and reserve power at the country's power stations, he raised the rate of utilization of installed capacity. As a result, Soviet power stations began to meet the demands of the grid more satisfactorily, but the risk of a major disaster increased.

In March 1986 (one month before Chernobyl), addressing the enlarged board of the ministry, Shcherbina felt that this achievement was worthy of commendation. Since at the time the deputy chairman of the Council of Ministers was in charge of fuel and energy matters in the government, his praise for Mayorets's work is understandable.

An experienced, ruthlessly demanding administrator, Shcherbina automatically transposed to matters of energy the management methods he had used in the gas industry, where he had for many years been a minister; he was rigid and not highly competent in energy-related and particularly nuclear matters. But this short, rather puny man had a lethally tenacious grip. Moreover, he had a striking ability to impose on nuclear power station builders his own deadlines for starting up reactors, while thinking nothing of accusing those same people, some time later, of failing to meet "the obligations they had assumed." Worse still, Shcherbina insisted on his own start-up deadlines, without taking any account of the time needed, from a technical standpoint, for the construction of nuclear power stations, the assembly of equipment, and start-up and verification procedures.

On 20 February 1986, while attending a meeting in the Kremlin of nuclear power station directors and the officials in charge of building the stations, I had observed a most curious pattern: when submitting their reports, the directors and construction heads spoke for 2 minutes each, while Shcherbina, who interrupted them constantly, spoke at least 35 to 40 minutes.

The most interesting statement was made by the senior construction officer of the Zaporozhiye nuclear power station, R. G. Khenokh, who plucked up the courage to state, in a deep bass voice—a bass voice, at such a meeting, was regarded as tactless—that the No. 3 reactor at Zaporozhiye could not, in the best of circumstances, be started up before August 1986 (it was actually started up on 30 December 1986) because equipment had been delivered late and the computing complex, on which assembly had only just begun, was not ready.

Shcherbina was indignant. "Well, that's just great! Here's a man who sets his own deadlines!" And, his voice rising to a shout, he went on, "Who gave you, Comrade Khenokh, the right to set your own deadlines instead of the ones set by the government?"

"Deadlines are dictated by the technology in use," the construction chief stubbornly replied.

Shcherbina interrupted him. "Forget it! Don't evade the issue! The government deadline is May 1986. Kindly start up in May!"

"But they won't finish delivering the special reinforcing bars until late in May," Khenokh rejoined.

"Have it delivered earlier," Shcherbina urged him, and, turning to Mayorets who was seated next to him, went on, "See, Anatoly Ivanovich, your construction chiefs are blaming everything on the absence of equipment and have been failing to meet deadlines."

"We'll take care of that, Boris Yevdokimovich," Mayorets promised.

"I fail to understand how a nuclear power station can be built and started up without equipment," Khenokh muttered. "After all, I'm not the one who supplies the equipment, it comes from industry through our client," and, thoroughly disgruntled, he sat down.

After the meeting, in the lobby of the Kremlin Palace, he told me,

"There you have our whole national tragedy in a nutshell. We ourselves tell lies, and we teach our subordinates to lie. Lies, even for a worthy cause, are still lies. And no good will come of it."

These remarks were made two months before the Chernobyl disaster.

In April 1982, I wrote an article about what I call "planning creep" in the nuclear construction industry. Planning creep occurs when, after one missed deadline for the start-up of a facility, new deadlines are repeatedly adopted without any attempt to understand what went wrong in the organization to cause the original failure. Such creeping delays often go on for many years, with a colossal increase in construction cost estimates. Although I submitted my article to one of the central newspapers, it was rejected. Here is a brief extract from it:

What are the causes of unrealistic planning in the nuclear construction industry and of the persistent failure, over whole decades, to meet deadlines? There are three of them:

1. The people who plan the timing of additional energy capacity, and manage the nuclear construction industry, are incompetent.
2. Unrealistic planning leads to planning creep, caused by incompetent evaluations.
3. The ministries in charge of machine building are unable to produce equipment of the right quality and in the required amounts for nuclear power stations.

Nuclear construction, like the operation of nuclear power stations, unquestionably demands a high degree of competence. As noted by then Foreign Minister Andrei A. Gromyko in the United Nations General Assembly on 2 November 1982, a major disaster at a nuclear power station with a breach of a reactor vessel is equivalent, in terms of some of its consequences, to the explosion of a megaton nuclear bomb. He must have sensed the coming of Chernobyl.

For that reason, those in charge of the construction and operation of nuclear power stations must have a thorough knowledge of that field. While this is obviously true of the operation of nuclear power stations (although here, too, there were a large number of violations, leading to Chernobyl), it might seem, at first sight, that when it comes to the *building* of nuclear power stations, competence in nuclear matters is perhaps superfluous. It could be argued that, after all, construction is construction: this goes here, that goes there, pour the concrete, what could be simpler? The simplicity is, however, only apparent. (It deceived both Shcherbina and Mayorets, who did not look before they leaped.)

From the very first load of concrete poured into its foundations, the task of constructing a nuclear reactor building is complicated by its future radioactivity and especially by the need for a timely start-up of the operating radioactive facilities, which is what nuclear power stations are. In other words, competence has a direct bearing on both the quality and the feasibility of the plan and also on the safety of a nuclear power station. However obvious these truths are, they unfortunately need to be repeated. Many of the senior people in the nuclear industry have no right to be where they are.

For example, immediately before Chernobyl, the central bureaucracy of the Ministry of Energy, including the minister and a number of his deputies, lacked competence in nuclear matters. The nuclear sector of energy-related construction was under the control of the sixty-year-old deputy minister A. N. Semyonov, by training and lengthy experience a builder of hydroelectric power stations, who had been appointed to his demanding post only three years previously. It was not until January 1987 that he was dismissed, on the basis of the results for 1986, when he failed to introduce new generating capacity on schedule.

Matters were no better with the operational management of existing nuclear power stations which, on the eve of the Chernobyl disaster, was in the hands of the All-Union Industrial Department for Nuclear Energy (known by the acronym Soyuzatomenergo). The chief was G. A. Veretennikov, a man who had never worked as an operator of a nuclear power station. He knew nothing about nuclear technology; and after working for fifteen years in the state planning agency, Gosplan, he decided to make the move to "hands-on work." (In July 1986, as a result of Chernobyl, he was expelled from the party and dismissed from his post.)

After the Chernobyl accident, Shcherbina, addressing representatives of the energy sector at a meeting of the full assembly of the Ministry of Energy in July 1986, stated, "All of these years you have been heading straight for Chernobyl!"

If so, then we should add that Shcherbina and Mayorets accelerated the march toward disaster.

NUCLEAR POWER
IN EAST AND WEST

In October 1979, an interesting article by Fred Olds, entitled "Outlook for Nuclear Power," was published in the journal *Power Engineering:*

> While a number of the OECD [Organization for Economic Cooperation and Development] countries are having trouble with their nuclear power programs, COMECON [Council for Mutual Economic Assistance] met and embarked on a joint nuclear power program that will add 150,000 MWe by 1990. This is more than a third of COMECON's present total generating capacity. Of the new capacity, 113,000 will be in the Soviet Union itself. . . .
>
> The nuclear program for these nations was set forth in late June as COMECON met on its 30th anniversary. Not surprisingly, there apparently is an oil spur behind the vigor of the nuclear plans. The Soviet Union exports oil to its East European economic allies and a million barrels [130,000 tons] a day to the West.* Last year [1978], however, the Soviet production was somewhat short of target, is apparently running short this year, and [is] predicted to fall short in 1980. Evidently the massive Siberian fields are proving difficult to exploit.
>
> Soviet Prime Minister Kosygin told COMECON that nuclear power is the solution to the energy problems. There is reported to be negotiation between Russia and West Germany for the sale of German hardware and technology to Russia, presumably to help the COMECON program along.† Earlier this year, Roumania signed a

---

*Here we must add that as of 1986 the Soviet Union annually exports to the West 336 million tons of conventional fuel equivalent—oil plus gas.
†The negotiations were broken off because the West German counterproposal contained certain unacceptable conditions.

$20 million licensing arrangement with Canada for four CANDU reactors, of 600 MWe capacity each. . . . According to one report, Cuba is slated to build one or more nuclear plants of Russian design. Supposedly, the design lacks containment and backup cooling features which are mandatory in Western designs.*

The Soviet Academy of Sciences has, not unexpectedly, assured the Russian people that Soviet reactors are safe, and that the Three Mile Island incident was exaggerated in the press. An eminent Russian nuclear scientist, Anatoly P. Alexandrov [president of the Academy of Sciences and director of the I. V. Kurchatov Institute for Atomic Energy], recently was interviewed by the London correspondent for the *Washington Star.* He is quoted as saying that failure to develop nuclear power poses great danger for mankind.

He regrets the United States using the Three Mile Island incident as a reason for slowing down nuclear power. Oil and gas are limited to 30–50 years, he said, and we must build nuclear plants in all parts of the world, otherwise wars will be fought, one day, over the remnants of oil and gas deposits. He sees the wars being fought between capitalist countries because the Soviet Union will have supplied itself with ample nuclear power. . . .

Two Blocs on Opposite Courses

In the industrialized world we see OECD and COMECON, both with a huge equity in oil, and with interestingly different approaches to their future energy supplies. COMECON has made nuclear power the spine of its program and downplays the possible contribution of solar and soft path options. The GDR [German Democratic Republic] looks for less than 2% from these sources. Environment takes a high priority, but productivity and a rising standard of living are placed before pristineness.

OECD members have a mix of nuclear energy programs, led by France . . . and Japan. Germany and the U.S. are in hold patterns. Canada is backing and filling for several reasons, and other members are generally slow. For OECD, the U.S. had been leading the way for some years in nuclear power deployment and probably in R&D expenditures. Then, in a short span of time, nuclear power went from the nation's highest priority to an energy option of last resort. Top

*Here Fred Olds is clearly wrong. The Cuban nuclear power stations being built from Soviet designs include both containment structures and auxiliary core cooling systems

priority is given to keeping the environment pristine, and zero risk is the first consideration in energy projects.

Thus, the lead countries of COMECON and OECD are on quite opposite nuclear power courses.

The positions are of course not diametrically opposed, especially with regard to safety improvements at nuclear power stations. Fred Olds is not entirely accurate at this point. Both sides devote the maximum attention to this question. There are unquestionably, however, differences of judgment about problems connected with the development of nuclear energy.

> In the United States, excessive criticism and obvious exaggeration of the hazards of nuclear power stations.
>
> In the Soviet Union, a total lack of criticism over a period of thirty-five years and obvious downplaying of the hazards of nuclear power stations for operational personnel and the environment.

A surprising phenomenon is the markedly conformist attitude of the Soviet public, which has unquestioningly believed the assurances of academicians and other highly placed incompetents. Surely that is why Chernobyl came at us like a bolt from the blue, and prompted radical rethinking on the part of so many people.

Not everyone, however, reacted in this way. Unfortunately, there are still plenty of credulous people who are quite willing to conform. After all, it is easier to believe what you are told than to subject it to sober scrutiny. To begin with, it is less trouble.

At the forty-first session of COMECON, held in Bucharest on 4 November 1986, seven years after the publication of Olds's article, once again we heard the confident claims of participants concerning the need for accelerated development of nuclear energy. The chairman of the Council of Ministers of the Soviet Union, N. I. Ryzhkov, addressing the session, said in particular:

> The tragedy at Chernobyl has not canceled out the prospects for nuclear energy on the basis of cooperation; in fact, by focusing our attention on the need for greater safety, it has increased the importance of nuclear energy as the only source guaranteeing reliable

supplies for the future. The socialist countries are becoming even more actively involved in international cooperation in this field, on the basis of our proposals to the International Atomic Energy Agency. Moreover, we shall build nuclear power stations for heating, while saving valuable and scarce organic fuel—gas and oil.

We must remember that nuclear power stations for heating will be built in the immediate vicinity of large cities, and pay special attention to the safety of such stations.

The vigor with which the development of nuclear energy is being advocated in the Soviet Union and the COMECON nations makes the lessons of Chernobyl all the more vital. We can learn from them, however, only by rigorously analyzing the causes, nature, and consequences of the disaster at the nuclear power station in the Byelorussian-Ukrainian Woodlands. I shall now try to do just that, by recording events as they occurred, hour by hour, day by day, before the accident and while it was actually happening.

2

THE ELEMENTS OF
THE TRAGEDY

FLYING OVER THE UKRAINE

ON THE EVE of the disaster, I was working as deputy director of the main industrial department in the Ministry of Energy dealing with the construction of nuclear power stations.

On 18 April 1986, I visited the Krymskaya plant, which was then being built, to see how construction and assembly were proceeding.

On 25 April 1986, at 4:50 P.M., eight and a half hours before the blast, I took off from Simferopol on an IL-86 bound for Moscow. Later I could recall no sense of foreboding or concern about anything in particular. On both landing and takeoff, there was a strong smell of jet engine fuel, which did bother me; but during the flight, the air was perfectly clean. The only thing that disturbed me slightly was the constant clanking of a poorly adjusted elevator carrying the cabin crew up and down with refreshments. They seemed flustered, and much of what they were doing was plainly redundant.

We flew over the Ukraine, where gardens were in bloom as far as the eye could see. Yet within another seven or eight hours that same land, the breadbasket of our country, would be plunged into a new era—one of calamity and nuclear contamination.

For the time being, however, I was gazing out the window at the earth. Kharkov floated by in a bluish haze. I remember regretting that Kiev was a little too far off the side. After all it was there, 80 miles from the capital of the Ukraine, that in the 1970s I had served as the deputy chief engineer at the No. 1 reactor unit of the Chernobyl nuclear power station. I used to live in the town of Pripyat, on Lenin Street, in the district exposed to the heaviest radiation after the explosion.

The Chernobyl nuclear power station is located in the eastern part of a large region known as the Byelorussian–Ukrainian Woodlands (Polyesye), on the banks of the Pripyat River, which flows into the Dnieper. The landscape is generally rather flat, with a slight slope toward the river and its tributaries.

The total length of the Pripyat before it reaches the Dnieper is 450 miles; it has a width of over 300 yards, a current flowing at about 3 feet a second, and an average discharge of 14,000 cubic feet (400 m^3) per second. The nuclear power station is located in a watershed of 41,000 square miles (106,000 km^2). And it is precisely from that area that the radioactivity was to enter the subsoil and be washed into the rivers by rain and melting snow.

I have such fond memories of the Pripyat River! Its water has a slightly brownish tinge, doubtless because it flows from the peat bogs of the Polyesye, and is densely saturated with fatty acids. The current is swift and strong, making swimming difficult. The skin on your body and arms becomes exceptionally taut, and when rubbed with the hand, your skin squeaks. I used to swim a lot in the river and also rowed on it, in school boats. After work I would go down to the boathouse, take out a single-seat scull, and spend an hour or two gliding over the smooth waters of this ancient river, as old as Russia herself. The river banks were quiet and sandy, covered with young pines; in the distance stood the railroad bridge, which echoed thunderously at eight o'clock each evening as the Khmelnitsky-Moscow passenger train passed over it.

And there was a feeling of primeval quiet and cleanliness. If you stop rowing and scoop up some brownish water in your hand, your palm immediately tends to contract on account of the fatty marsh acids in the water. It was those same acids that, after the explosion and the release of radioactive substances, served as efficient coagulators, transporting radioactive particles and fission fragments.

But let me return to the Chernobyl plant. Its site is not without importance.

The aquifer on which local users, including industry, rely for water lies at a depth of 33–49 feet (10–15 m) below the level of the Pripyat River and is separated from the quarternary deposits by layers of virtually impermeable clay. Thus the radioactivity, once it reached that depth, was transported horizontally by the groundwater.

In the region of the Byelorussian–Ukrainian Woodlands, population density is generally low. Before the construction of the Chernobyl nuclear power station got under way, there were roughly 180 people per square mile (70 per km²). By the time of the disaster, 110,000 people were living within an 18.5-mile (30-km) radius of the plant—almost half of them in the town of Pripyat, situated west of the 1.9-mile (3-km) safety zone around the plant, and 13,000 in the regional center, Chernobyl, located 11 miles (18 km) to the southeast of the plant.

I had fond memories of the settlement in which the nuclear workers were housed. I saw it built almost from the ground up. By the time I had left to work in Moscow, three of its districts were already inhabited. It was comfortable, convenient, and very clean. Visitors were often heard to say, "What a charming place Pripyat is!" Many retired persons, finding Pripyat an attractive, well-designed town in beautiful natural surroundings, went to considerable trouble to acquire residence permits; some of them pressed their case in government agencies and even in the courts.

Not so long ago, on 25 March 1986, I went to Pripyat to see how construction of the No. 5 reactor unit was proceeding. Again, I found that same intoxicatingly clean fresh air, the same quiet and coziness, of what was then no longer a settlement, but a town with 50,000 inhabitants.

By now the plane had left Kiev and the Chernobyl nuclear power station behind, to the northwest. My reminiscences faded, and I was left contemplating the spacious interior of the plane, with its two aisles and three sections of half-empty seats. I somehow felt that I was inside a huge barn, and that if I shouted, my greeting would echo back to me. Next to where I was sitting the elevator continued to rumble and clank on its way up and down. It almost seemed I was not in a plane, but riding in an old-fashioned carriage along a blue, cobble-stoned road, with milk churns rattling in the trunk.

I arrived from Moscow's Vnukovo airport by nine o'clock that evening. Five hours before the blast.

A TEST PROGRAM
BYPASSING THE SAFETY RULES

At Chernobyl on that same day, they were preparing to shut down No. 4 reactor unit for scheduled maintenance. According to the program drawn up by the chief engineer, N. M. Fomin, while the unit was shut down for maintenance, tests were to be carried out bypassing the reactor safety systems; the equipment in the nuclear power station was to be fully de-energized, with electricity being generated by the kinetic energy of the turbine's rotor blades.

In actual fact, such a test had been proposed to numerous nuclear power stations, but they all refused to take part, on account of the risks involved. However, the people in charge at Chernobyl agreed.

What was the point of the experiment?

If all power is cut off to the equipment in a nuclear power station, as can happen in normal operations, all machinery stops, including the pumps that feed cooling water through the reactor core. The resulting meltdown of the core is a nuclear accident of the utmost gravity.

As electricity must be generated by any means available in such circumstances, the experiment using the residual inert force of the turbine is an attempt to provide a solution. As long as the turbine blades continue to spin, electricity is generated. It can and must be used in critical situations.

Similar tests—only with the reactor safety systems turned on— had previously been conducted at other nuclear power stations. And everything had gone well. Indeed, I had taken part in them myself on occasion.

The schedule for operations of this sort is usually worked out in advance and coordinated with the chief designer of the reactor, the general project manager of the nuclear power station, and Gosatomenergonadzor, the USSR State Committee on Operational Safety in the Nuclear Power Industry (henceforth referred to as the Nuclear Safety Committee). The schedule for such events has to provide for reserve power supply, for the duration of the experiment, to those top-priority systems and equipment that cannot tolerate interruptions lasting more than a split second or a few seconds. The cutoff of energy for the needs of the station itself during the tests is hypothetically assumed, but does not actually take place.

In such circumstances, power for the needs of the station is drawn from the system through the operational and start-up or standby transformers and also through the independent power provided by two standby diesel generators.

Nuclear safety during the tests requires the functioning of both the emergency reactor safety system, which is triggered when certain limits are exceeded and inserts the control rods into the reactor core, and of the emergency core cooling system.

Provided that the established rules were complied with, and auxiliary safety measures taken, such tests at operational nuclear power stations were not prohibited.

I must emphasize, however, that tests involving the inert spinning force of the generator blades should be conducted only after the reactor's emergency power reduction (or "scram") system, designated in Russian by the letters AZ, has been activated by pressing the button for that function. Until then the reactor must be kept in a stable, controllable regime, with the reactivity reserve stipulated in the regulations.

The program approved by the chief engineer at the Chernobyl nuclear power station, N. M. Fomin, failed to meet any of these requirements.

Before I proceed, here are some explanations the general reader might find useful.

RBMK, the designation of the reactors in use at Chernobyl, stands for "high-power channel-type reactor." Put simply, the core of an RBMK reactor consists of a cylinder about 46 feet (14 m) in diameter and 23 feet (7 m) tall. Inside, it is tightly packed with graphite columns, each of which contains a tubular channel. The nuclear fuel bundles are loaded into these channels, thus forming the "fuel assemblies." The tubular openings evenly distributed all over the top end of the core cylinder receive the control rods, which absorb neutrons. When all the rods are lowered within the core, the reactor is shut down. As the rods are withdrawn, the chain reaction of nuclear fission begins and the power of the reactor increases. The higher the rods are withdrawn, the greater the power of the reactor.

When a reactor has just been loaded with fresh fuel, its reserve reactivity (in other words, its ability to increase neutron power) exceeds the ability of the absorbent rods to suppress the chain reaction. In such circumstances, parts of the fuel bundles are withdrawn, and stationary absorber rods known as auxiliary absorbers are inserted in their place to assist the movable rods. As the uranium is gradually burned up, these auxiliary rods are removed and replaced by nuclear fuel.

There is, however, one rule that absolutely must be followed: as the fuel is consumed, at least 28 to 30 absorber rods must still be inserted in the core (after Chernobyl the number was raised to 72). This is because the power-generating capacity of the fuel may at any time exceed the ability of the control rods to absorb neutrons and thus slow down the chain reaction in the core. Those 28–30 rods located in the zone of high differential worth—that is, where they have the greatest effect on the core's ability to generate power—constitute the operational excess reactivity margin. In other words, at all phases of the operation of a reactor, its capacity for nuclear power generation must not exceed the ability of the absorber rods to suppress the chain reaction.

No. 4 unit at the Chernobyl nuclear power station was brought on-line in December 1983. At the time of the shutdown of the unit for scheduled maintenance, due to take place on 25 April 1986, the reactor core contained 1,659 fuel assemblies containing about 197 tons (200 metric tons) of uranium dioxide, one auxiliary absorber, loaded into a fuel channel, and one empty fuel channel. Most (in fact 75%) of the heat-emitting assemblies were bundles from the original fuel, with a burn-depth close to the maximum levels—a indicates very high levels of long-lived radionuclides in the core.

The tests scheduled for 25 April 1986 had been conducted previously at the plant. They had shown that tension on the generator's bus-bars, or connecting conductors, fell long before the generator blades exhausted their kinetic energy in the inertial rundown. The scheduled tests included the use of a special device to regulate the generator's magnetic field and thereby eliminate this shortcoming.

One is naturally inclined to ask why previous tests had occurred without an emergency. The answer is simple: the reactor was in a stable, controlled state, with the entire range of safety systems working normally.

But this time the program was poorly prepared and its safety-related measures had been drawn up as a pure formality. They merely stipulated that all switching operations carried out during the experiments were to have the permission of the plant shift foreman; and that in case of emergency, the staff was to act in accordance with local instructions. Before the start of the tests, the supervisor of electrical aspects of the experiment—Gennady Petrovich Metlenko, an electrical engineer who was not a specialist in reactor plants—would advise the security officer on duty accordingly.

Besides making essentially no provision for supplementary safety measures, the program stipulated that the emergency core cooling system (ECCS) should be switched off. Thus throughout the intended duration of the tests—about 4 hours—the safety of the reactor was greatly reduced.

As safety was neglected in the test program, the personnel were not ready for the tests and were quite unaware of its possible dangers.

Moreover, as we shall see, plant personnel did not even comply with the terms of the program, thereby further exacerbating the situation. The operators were not fully aware that the RBMK reactor has a series of positive reactivity coefficients, which in certain cases take effect simultaneously and lead to what we call a "positive shutdown," or an explosion. And in actual fact, that same power surge played a fateful role.

THE DECISION MAKERS

But let us now take another look at the test program, and try to understand why it had not been coordinated with those higher organizations that, like the senior managers at Chernobyl, bore responsibility for the nuclear safety not only of the plant itself but also of the entire country.

In January 1986, the plant director, V. P. Bryukhanov, sent the program to the general project manager at the Gidroproyekt institute and to the Nuclear Safety Committee. No reply was received.

Neither senior management at the Chernobyl nuclear power station, nor the operational department of Soyuzatomenergo found such a development particularly troubling. Gidroproyekt and the Nuclear Safety Committee did not seem to care, either.

Certain far-reaching conclusions can already be drawn: irresponsibility and carelessness at these state agencies had reached the point where they all found it possible to say and do nothing—although Gidroproyekt, which was in charge of the plant's overall design, Soyuzatomenergo, its principal client, and the Nuclear Safety Committee all had the power to impose sanctions. Indeed, it was their direct duty to do so. Still, specific individuals within those organizations were responsible for such matters. Who were they? Were they capable of fulfilling the responsibility entrusted to them?

At Gidroproyekt, the man in charge of safety at nuclear power stations was V. S. Konviz, an experienced designer of hydroelectric power stations, who had written a dissertation on hydrotechnical facilities. For many years, from 1972 to 1982, he had been the director of the unit designing nuclear power stations; in 1983 he had been placed in charge of safety at nuclear power stations. When he began designing nuclear power stations in the 1970s, Konviz had only the vaguest idea what a nuclear reactor was; he studied nuclear physics out of a high school textbook and recruited experts on hydrotechnology to work on nuclear plant design. His case, at least, is perfectly clear. Such a person could not have anticipated the possibility for disaster that had been built into the program and, indeed, into the reactor itself.

The reader is perhaps wondering why he chose to work in a field with which he was clearly unfamiliar. The answer is that it brought him prestige, money, convenience. The same was true of Mayorets, Shcherbina, and many others.

At Soyuzatomenergo—the department of the Ministry of Energy which was in practice responsible for all the actions of operational staff—the director was G. A. Veretennikov, who had never himself done any operational work at nuclear power stations. From 1970 to 1982, he had worked at the State Planning Committee, Gosplan, initially as a principal specialist and then as head of a unit in the Department of Energy and Electrification. He dealt with questions involving the planning of supplies of equipment for nuclear power stations. Supplies, for various reasons, were in a sorry state. Year after year, up to 50 percent of the planned hardware was not being delivered. Veretennikov was frequently sick; it was said that he had trouble with his head, suffering from spasms of the cerebral blood vessels. Nonetheless, his personal ambitions were highly developed. In 1982, bringing all his connections into play, he took over the vacant combined post of deputy minister and director of Soyuzatomenergo. It was more than he could handle, even from the purely physical point of view. Then came another succession of spasms of the cerebral blood vessels, fainting spells, and lengthy stays in the Kremlin hospital. Yu. A. Izmailov, a veteran of Glavatomenergo, the central directorate for nuclear power, used to joke about it: "Under Veretennikov it was practically impossible for us to find anyone in the central directorate who knew much about reactors and nuclear physics. At the same time, however, the bookkeeping, supply, and planning department grew to an incredible size."

In 1984 the post of deputy minister was eliminated, and Veretennikov became simply the director of Soyuzatomenergo. This was more of a blow to him than the Chernobyl explosion. His fainting spells became more frequent, and he was once again admitted to hospital.

Not long before Chernobyl, the director of the production branch of Soyuzatomenergo, Y. S. Ivanov, sought to explain away the increasingly frequent breakdowns at nuclear power stations:

"Not one nuclear power station complies fully with the operating rules. And it's impossible for them to. Corrective action has to be taken all the time, in the light of actual experience."

It took the nuclear disaster at Chernobyl to have Veretennikov expelled from the party and from his post as head of Soyuzatomenergo. It seems a pity that the only way our bureaucrats can be removed from office is by being blasted out of their comfortable armchairs.

The Nuclear Safety Committee was staffed by some well-trained and experienced people, under the committee chairman Ye. V. Kulov, an experienced nuclear physicist, who had previously worked for many years on nuclear reactors for the Ministry of Medium Machine Building.* Curiously, however, even Kulov failed to react to the poorly drafted test program from Chernobyl. One wonders why that was so: after all, decree 409, issued by the Council of Ministers on 4 May 1984, stipulated that the Nuclear Safety Committee was required to perform the following functions: "Monitoring, on behalf of the State, of compliance by all ministries, departments, enterprises, organizations, institutions, and officials of the established rules, norms, and instructions for nuclear and technical safety in the design, installation, and operation of nuclear power plants." Moreover, paragraph *(zh)* authorizes the committee to "take appropriate measures, including closing down nuclear power facilities if safety rules and norms are violated, if there is evidence of defects in equipment, or of incompetence on the part of the personnel, and also in other cases where the operation of these facilities is jeopardized."

At one of the meetings held in 1984, Kulov, who had recently been appointed chairman of the Nuclear Safety Committee, explained to an audience of nuclear plant personnel how they were to function: "Do not imagine that I am going to do your jobs for you. Figuratively speaking, I am a policeman. My job is to forbid and cancel out any wrong moves on your part." Unfortunately, even as

*This secretive ministry is in charge of reactor design and the nuclear fuel cycle, with subdivisions involved in plutonium production, reprocessing of spent fuel, and the manufacture of nuclear weapons.—Ed.

a policeman, Kulov did not do a particularly good job in connection with Chernobyl.

What prevented him from halting operations at No. 4 reactor? The test program, after all, could not stand up to criticism.

And what prevented Gidroproyekt and Soyuzatomenergo from acting?

It was almost as if they had conspired not to intervene. Why? The fact is that there was a conspiracy of silence. Mishaps were never publicized; and as nobody knew about them, nobody could learn from them. For thirty-five years people did not notify each other about accidents at nuclear power stations, and nobody applied the experience of such accidents to their work. It was as if no accidents had taken place at all: everything was safe and reliable. No account was taken of the wise words of Abutalib: "If a man fires at the past from a pistol, the future will fire at him from a cannon." For those involved in the nuclear power industry, I would paraphrase as follows: "The future will blast them with a nuclear disaster from a reactor."

Here I should add another detail which was not mentioned in any of the technical reports on Chernobyl: the regime for the inertial rundown of the generator used in one of the subsystems of the fast emergency core cooling system (ECCS) had been planned previously, and not only was embodied in the test program but had also been technically prepared. Two weeks before the experiment, an MPA button (the initials of the Russian for "maximum design-basis accident") had been added to the control panel of No. 4 reactor; the signal it produced when pressed went only to the secondary electrical circuits, bypassing the monitoring and measurement instruments and the pumps. In other words, the signal from that button was a fake, as it bypassed all the most important parameters and trip mechanisms of the nuclear reactor. A serious mistake!

As failure of the 31.5-inch (800-mm) suction or pressure header in the reinforced leaktight compartment is deemed to constitute the start of a maximum design-basis accident, the parameters activating the emergency power reduction system (AZ) and the ECCS system were: a reduction of pressure in the intake line of the main circulation pumps; a reduction of the pressure gradient from the lower water communication lines to the separator drums; and an increase in pressure in the reinforced leaktight compartment.

Normally the emergency power reduction system (AZ) switches itself on when these parameters have been reached. All 211 absorbent rods are lowered; cooling water is fed from the ECCS tanks; the emergency feedwater pumps switch themselves on; and the standby diesel generators start up. Pumps also start delivering an emergency supply of water from the pressure suppression pool to the reactor. Thus, there are plenty of emergency protection systems, if they are kept operational and function when needed.

The crucial point here is that all of these protective systems should have been connected to the MPA button. Most unfortunately, however, they were switched off, as it was feared that cold water might reach the reactor and produce a heat shock. This utterly insignificant notion apparently mesmerized both the senior officials at the plant (Bryukhanov, Fomin, and Dyatlov) and the higher organizations in Moscow. In this way a sacrosanct rule of nuclear technology was violated. After all, if the maximum design-basis accident was allowed for in the design, it could happen at any time. Who authorized the removal of all the protective systems stipulated in the design and in the rules for nuclear safety? No authorization was given. They simply took it upon themselves.

One also wonders why such irresponsibility on the part of the Nuclear Safety Committee, Gidroproyekt, and Soyuzatomenergo did not alarm Bryukhanov and Fomin, director and chief engineer at the Chernobyl plant, respectively. After all, it is impossible to work on the basis of an uncoordinated program. What kind of people—and what kind of experts—were Bryukhanov and Fomin?

I made the acquaintance of Viktor Petrovich Bryukhanov in the winter of 1971 when I went to the construction site for the nuclear power station, to the village of Pripyat, straight from the Moscow clinic where I had been receiving treatment for radiation sickness. Though I still felt ill, I could walk and decided that going to work would speed up my recovery.

Having signed a form confirming that I was being discharged voluntarily, I took a train and by morning was in Kiev. From there a two-hour taxi ride took me to Pripyat. Several times on the road

I felt nauseated, confused, and dizzy, but I was anxious to make a start at the job to which I had been assigned shortly before my illness.

Fifteen years later, after the disaster in No. 4 reactor, firefighters with lethal doses of radiation and injured operational staff from the plant were taken to that same Clinic No. 6, where I had been a patient. However, in those days the foundations of the main building at Chernobyl were still being excavated, and there was nothing at all on the site of the future nuclear power station—just a few young pine trees in the blissfully clean air. If only they could have known where not to build their foundations!

As we approached Pripyat, I had already noticed the sandy soil of the hills, covered with low trees, with frequent flashes of clean yellow sand against a background of dark green moss. There was no snow. In some places, the warmth of the sun had brought forth green grass. All about there was a feeling of peace and pristine purity.

"This is poor-quality land, but very old," said the taxi driver. "Here, in Chernobyl, Prince Svyatoslav chose himself a bride—quite a lively girl, apparently. This little town is more than a thousand years old. And it's still here, it's not dead yet."

The winter day in Pripyat was sunny and warm. I have often remembered it like that in later years. It was still winter, but the air smelled of spring. The taxi driver stopped near a long wooden hut, where the management of the future nuclear power station and the senior construction officials were temporarily housed.

I went into the hut. The floor sagged and creaked under my feet. There was the director's office, quite a small room, about 65 square feet (6 m²). The chief engineer, Mikhail Petrovich Alekseyev, who was later appointed deputy chairman of the Nuclear Safety Committee, had a similar office. As a result of the Chernobyl disaster, he was to be severely reprimanded and his personnel record gravely tarnished. But all that was much later.

When I entered, Bryukhanov, a short man with fair curly hair and a tanned lined face, rose to his feet. Smiling awkwardly, he shook my hand. His whole appearance suggested that he was a gentle, pliable individual. That first impression was later confirmed, but certain other facets of his character also came to light—in particular, an inner stubbornness, accompanied by a poor knowledge of people; this trait led him to rely on employees who, though worldly-wise, were sometimes lacking in integrity. In those days, Bryukhanov was quite young—only thirty-six. By both training and experience he specialized in turbines. He distinguished himself as a student at the energy institute. He progressed well at the Slavyanskaya coal-fired power station, where he performed particularly well during the start-up phase. Sometimes he would stay at work for days on end, finding prompt and intelligent solutions as problems arose. In general, as I later found by working side by side with him for several years, he was a good, hardworking engineer with a quick mind; but unfortunately he was not a nuclear power expert. And that, when all is said and done, proves to be the most important consideration of all, as Chernobyl made clear. People who work at nuclear power stations just have to be experts in nuclear power.

The deputy minister of energy of the Ukraine, who was also in charge of the Slavyanskaya plant, noticed Bryukhanov and nominated him for the post at Chernobyl.

Bryukhanov had a narrow range of interests, was not particularly well read, and had a rather limited general culture, all this accounting to some extent, I later felt, for his tendency to surround himself with people who were not themselves topnotch.

At the time, however, in 1971, when I introduced myself, he said with evident pleasure, "Ah, Medvedev! We have been expecting you. You can get to work right away." He left the office and summoned the chief engineer.

Mikhail Petrovich Alekseyev, who then entered the room, had already been working at Chernobyl for several months, having gone there from the Byeloyarsk nuclear power station, where he had been the deputy chief engineer for the third reactor unit, which existed only on paper. Alekseyev had no experience of nuclear operations; before Byeloyarsk he had spent twenty years at thermal power stations. And, as soon became evident, he was determined to move up to Moscow, where he went about three months after I started work at Chernobyl. I have already referred to the punishment he received for his part in the 1986 disaster. His superior in Moscow, the chairman of the Nuclear Safety Committee, Ye. V. Kulov, was penalized much more harshly: he was dismissed and expelled from the party. Bryukhanov suffered the same fate before being brought to trial.

All that, however, was to happen fifteen years later, by which time some important events had occurred, particularly with regard to personnel policy in the nuclear power industry. I contend that it was this policy, which Bryukhanov also pursued, that led to the events of 26 April 1986.

Soon after starting work at Chernobyl (having already worked for many years as shift foreman at another nuclear power station), I set about hiring people to work in the various sections and units at the plant. I would present Bryukhanov with a list of candidates who had abundant experience of work at nuclear power stations. As a rule, he did not directly reject them, but he also did not recruit them. Instead, he would, in a leisurely manner, proceed to nominate candidates from thermal power stations, sometimes even appointing them himself. In so doing, he would argue that the kind of candidates best suited for work at nuclear power stations were experienced power station people with a thorough knowledge of powerful turbine systems, distribution systems, and power transmission lines.

It was possible only with great difficulty, by going over Bryukhanov's head and first securing the support of Glavatomenergo, to staff the reactor and specialized chemical units with the needed specialists. Bryukhanov recruited people for the turbine and electrical departments. Nikolai Maksimovich Fomin and Taras Grigoryevich Plokhy came to work at Chernobyl toward the end of 1972. Bryukhanov nominated Fomin as chief of the electrical unit, and Plokhy as deputy chief of the turbine unit. Both men were sponsored directly by Bryukhanov. Fomin, who had trained and worked as an electrical engineer, was sent to Chernobyl from the Zaporozhiye State regional electric power station, a thermal facility, and worked before that in the Poltava power grid. I mention these two names because fifteen years later they were to be associated with two major disasters, at Balakovo and Chernobyl.

As deputy chief engineer for operations, I talked to Fomin and warned him that a nuclear power station was a radioactive and extremely complex facility. Had he thought things over carefully before moving out of the electrical unit of the Zaporozhiye thermal plant?

Fomin had a nice smile, revealing a full set of gleaming white teeth. He was evidently aware of this and kept smiling practically nonstop, whether or not it was appropriate to do so. With a knowing smile he replied that a nuclear power station was a prestigious ultramodern place to work; and that, in any case, you didn't have to be a genius to run one—there was really nothing to it.

He spoke in a pleasant and vigorous baritone, which climbed toward the alto whenever he was excited. He had a square, angular figure and dark sparkling eyes which stared right through you. In his work he was precise, assiduous, demanding, impulsive, and ambitious, and he could certainly hold a grudge. His gait and movements were brusque. It seemed that he was inwardly coiled up like a spring, ready to be unleashed at any moment. I have dwelt on him in such detail because he was destined to play a part in history, as a sort of nuclear antihero, whose name was to be associated, starting on 26 April 1986, with one of the most terrible nuclear power station disasters.

Taras Grigoryevich Plokhy, on the other hand, was sluggish but thorough—a phlegmatic type, with a tedious, slow way of speaking, but meticulous, obstinate, and hardworking. On meeting him for the first time, one would be inclined to say he was vague and something of a marshmallow, were it not for his methodical and intensive approach to work. Moreover, his close relationship to Bryukhanov, with whom he had worked at the Slavyanskaya plant, concealed a great many things. In the reflected light of that friendship, he struck many people as more important and vigorous than he really was.

After I had left Pripyat to work in Moscow, Bryukhanov began to arrange for Plokhy and Fomin to be promoted to the senior echelons at the Chernobyl plant. Plokhy led the way, eventually becoming the deputy chief engineer for operations and then chief engineer. He did not stay long in that post, being appointed, on the proposal of Bryukhanov, chief engineer at the Balakovo nuclear power station, then under construction, though he was unfamiliar with the design of the pressurized water reactor at that plant. During start-up procedures at Balakovo in June 1985—as a result of carelessness and slipshod work on the part of the operational staff under his command, and also of a serious breach of the safety rules—there was an accident in which fourteen men were boiled alive. The bodies were hauled out from the annular compartments surrounding the reactor well to the emergency lock and laid at the feet of the incompetent chief engineer, who stood there, his face a deathly white.

Meanwhile, back at the Chernobyl plant, Bryukhanov continued to advance the career of Fomin, who rapidly rose through the posts of deputy chief engineer for assembly and operations and soon replaced Plokhy as chief engineer. It is worth noting that the Ministry of Energy did not support Fomin's nomination. V. K. Bronnikov, who had experience in nuclear power stations, was a candidate for the post. Kiev, however, refused to take him, claiming that he was merely a run-of-the-mill technician, whereas Fomin—a tough, demanding leader—was the man they needed. Moscow conceded. Fomin was endorsed by the nuclear section of the Central Committee of the Communist Party, and the matter was resolved. The price of that concession is well known. At this point there was clearly every reason to pause for a moment, to take stock and ponder the implications of what had happened at Balakovo, and call for greater vigilance.

Late in 1985, Fomin suffered spinal injuries in a car crash, which left him paralyzed for some time and extremely depressed about his future. His robust organism overcame this setback, however; he recovered and returned to work on 25 March 1986, one month before the Chernobyl explosion. I happened to be in Pripyat at the time, inspecting No. 5 unit which was under construction. Progress was being impeded by a lack of design documentation and specialized hardware. I saw Fomin at a meeting we had convened specially about No. 5 unit. He was a shadow of his former self. His whole being had clearly slowed down and bore the imprint of the pain he had endured. The effects of the car crash were still plainly visible.

"Perhaps you ought to take a few more months off and get better," I suggested. "Those were pretty serious injuries."

"No, it's all right," he replied brusquely and with a rather forced laugh. His eyes had the same tense, malicious, and feverish look I had noticed fifteen years before. "Got to get on with the job."

I still felt that Fomin was not well, and that his condition was dangerous not only for him personally but also for the four reactor units under his operational control. This worried me, so I decided to share my apprehensions with Bryukhanov; but he, too, tried to reassure me: "I don't think it's particularly serious. He's made a good recovery. He'll get back to normal faster by working."*

Although I found such confidence troublesome, I did not press my objections. After all, was it any of my business? Maybe he really did feel well. In any case, since I was then working on nuclear plant construction, I had nothing to do with operational questions and therefore was unable to do anything about removing Fomin from his post or finding a temporary replacement for him. After all, he had been cleared for a return to work by experienced doctors, specialists who knew what they were doing. Nonetheless, my doubts persisted, and I felt compelled to draw Bryukhanov's attention to what I regarded as Fomin's ill health. But Bryukhanov reassured me. Then we talked a while. Bryukhanov complained that there had been numerous leaks at the plant, especially in the drainage channels and the air vents; also, the metal reinforcing bars were not holding up. The total volume of leaks amounted to an almost constant 1,765 cubic feet (50 m³) an hour of radioactive water, and it was hardly possible for the steam extraction units to keep up with the amounts to be reprocessed. There was a lot of radioactive waste. He told me that he was already very tired and would like to take a job somewhere else.

He had just returned from Moscow where he had been a delegate to the twenty-second congress of the Communist Party.

THE EVENTS OF 25 APRIL 1986

But what exactly happened at No. 4 reactor of the Chernobyl nuclear power station on 25 April, while I was still at the Krymskaya plant and later aboard an IL-86 on my way to Moscow?

---

*V. S. Smagin had this to say about Fomin: "He was a good worker, quick to take offense, pushy, vain, vindictive, spiteful, though he was occasionally fair. His voice was a usually pleasant baritone but sometimes shot up an octave when he got excited."

At 1 P.M. on the afternoon of 25 April 1986, the operational staff started to lower the power of No. 4 reactor, which was functioning within the normal parameters, at 3,000 megawatts (thermal).

Power was being reduced on orders from Dyatlov, the deputy chief engineer in charge of operations for No. 3 and No. 4 reactors (phase 2 of the Chernobyl plant), who had prepared No. 4 reactor for the implementation of the program approved by Fomin.

At 1:05 P.M., No. 7 turbogenerator was disconnected from the grid while the thermal output of the reactor was 1,600 megawatts (thermal). Electricity for the unit's own needs (four main circulation pumps, two electric feed pumps, and other equipment) was transferred to the bus-bars of No. 8 turbogenerator, which was still running and was to be used for the tests devised by Fomin.

At 2 P.M., in keeping with the experimental program, the emergency core cooling system was disconnected from the multiple forced circulation loop. This was one of Fomin's most severe and fatal mistakes. And it was done deliberately, so as to prevent the cold water from the ECCS tanks from entering the hot reactor, causing a heat shock. When the prompt neutron power surge later started and the main circulation pumps were disconnected, thus leaving the reactor without cooling water, 12,360 cubic feet (350 m^3) of emergency water from the ECCS tanks might just have saved the day, by suppressing the reactivity void coefficient. This, the most damaging factor of all, refers to the extent to which steam influences the nuclear reaction: in the RBMK, as in other graphite-moderated reactors, the formation of steam tends to enhance the chain reaction.

Who knows what might have come of it? But a person determined to make his mark as a leader, to distinguish himself in a prestigious sector and prove that a nuclear reactor, unlike a transformer, can function without cooling—such a person is capable of anything.

It is hard to imagine what secret schemes were going through Fomin's mind in those fateful hours; but only a man with no understanding of the processes of neutron physics inside a nuclear reactor could possibly have switched off the emergency core cooling system, which could in the critical seconds have prevented the blast by sharply reducing steam content in the core. At the very least, he must have had extravagant confidence in his own powers.

Nonetheless, it was done—and done, I repeat, quite deliberately. Evidently the same self-confidence—contrary to the laws of physics—must have clouded the judgment of the deputy chief engineer for operations, Dyatlov, and the entire staff in charge of No. 4 reactor. Otherwise, one of them—even just one—should have come to his senses when the ECCS was switched off, and shouted out, "Stop! What do you people think you're doing? Just take a look around. We're only a short distance from some ancient cities—Kiev, Chernigov, Chernobyl—and from the most fertile soil in the country, the blooming orchards of the Ukraine and Byelorussia. Mothers are giving birth in the Pripyat maternity clinic at this very moment! Their babies must come into a clean world! Come to your senses!"

No one shouted out, and no one came to their senses. The ECCS was calmly turned off; the gate valves on the line feeding water to the reactor were de-energized beforehand and put under lock and key, so that if the need arose they could not be opened even by hand. Otherwise, they might have been thoughtlessly opened, and 12,360 cubic feet (350 m^3) of cold water would have slammed into the red-hot reactor. The fact is, however, that in a maximum design-basis accident cold water would in any case enter the core. It was just a question of choosing the lesser of two evils. It is better to feed cold water into a hot reactor than to leave a red-hot core without water. As the Russian saying goes, "There's no point in crying over a man's hair when his head has been chopped off." The water from the ECCS arrives exactly when it needs to, and a heat shock is incomparably less serious than an explosion.

Psychologically, analysis of the situation is not easy. Admittedly, the conformist mentality of the operators, who had lost the habit of thinking for themselves, and the careless, slipshod attitudes that had crept in, had become well established in nuclear power station management; indeed, they had become the norm. Another factor was lack of respect for the nuclear reactor, which, in the minds of the operational staff, was scarcely more complicated than an ordinary samovar. They had forgotten the golden rule of workers in dangerous factories: "Remember! Wrong moves cause explosions!" With an electrical engineer in charge, there was also an electro-technical bias in their thinking; moreover, that same chief engineer was still under the psychological effects of serious spinal and cerebral trauma. There was also unquestionably a failure on the part of the psychiatric unit in the medical service of the Chernobyl nuclear power station, which was supposed to monitor the mental health of operators and management at nuclear plants and, when necessary, to relieve them of their duties.

Thus, the emergency core cooling system was disconnected deliberately, in order to prevent a heat shock in the reactor when the MPA button (maximum design-basis accident) was pressed. Apparently Dyatlov and the operators were confident the reactor would not fail them. Over confidence? Most certainly. It is clear that the operational staff did not fully understand the physics of the reactor and had not expected the situation to get completely out of hand. I think that the ten years of relatively successful operations at the Chernobyl plant had also induced a less vigilant state of mind. And even the ominous warning of September 1982, when a partial core melt occurred in No. 1 reactor of the Chernobyl plant, failed to teach them a lesson. In actual fact, it could not have taught them anything. For many years, accidents at nuclear power stations had been kept secret, although the operators themselves occasionally heard about them from each other. However, they did not take them seriously enough, doubtless feeling that if their superiors were keeping quiet about such things, that was good enough for them. Moreover, accidents were already being viewed as unpleasant, but inevitable, corollaries of nuclear technology.

Over a period of decades, the operators' confidence gradually became greatly exaggerated and made it possible for the laws of nuclear physics and the safety rules to be completely overlooked.

The start of the experiment was, however, postponed. At the request of the load dispatcher in Kiev at 2 P.M. on 25 April 1986, the removal of the unit from the grid was delayed. In a clear breach of the safety rules, No. 4 reactor was kept operating while the emergency core cooling system was switched off, although the official reason for doing so was the existence of the MPA button and the fact that the protective systems had been bypassed on the grounds that pressing that button would send cold water into the hot reactor. Of course, neutralizing the protective systems in such circumstances was a criminally irresponsible act.

At 11:10 P.M., when the shift foreman at No. 4 reactor was Yuri Tregub, the reduction of power was resumed.

At midnight, Yuri Tregub was relieved by Aleksandr Akimov, and his senior reactor control engineer was relieved by Leonid Toptunov.

At this point a question arises: If the experiment had been conducted during Tregub's shift, would the explosion have taken place? I do not believe it would. The reactor was in a stable, manageable condition; the operational reactivity reserve was more than 28 absorbent rods, and the power level was 1,700 megawatts (thermal). However, the experiment could just as easily have ended in an explosion during that shift if, when switching off the local automatic control system, the senior reactor control engineer on Tregub's shift had made the same mistake as Toptunov and then, having made it, had begun to climb out of the "iodine well." It is hard to say what would have happened; but it is to be hoped that the senior reactor control engineer on Tregub's shift would have done a more professional job than Leonid Toptunov and been more adamant in resisting pressure from his superior. The human factor clearly played a role here.

But developments proceeded as programmed by Fate. The apparent reprieve granted by the Kiev load dispatcher, when he shifted the tests from 2 P.M. on 25 April to 1:23 A.M. on 26 April, actually led straight to the explosion.

In keeping with the test program, the inertial rundown of the rotor blades of the generator, which was to supply its own needs, was supposed to occur at a power level of 700–1,000 megawatts (thermal). Such a rundown should really have taken place when the reactor was being shut down, because in a maximum design-basis accident the reactor's emergency power reduction system, on the basis of the five emergency parameters, is supposed to be triggered and to shut down the reactor. Yet a different, disastrously hazardous course was followed: the generator was allowed to run down while the reactor was still functioning. Why such a dangerous regime was chosen remains a mystery. One can only assume that Fomin wanted a totally pure experiment.

It is possible to move all the absorber rods simultaneously or in groups. With one of these local systems disconnected, as provided for in the operational rules for nuclear reactors at low power, the senior reactor control engineer, Leonid Toptunov, was unable to eliminate the imbalance in the measurement functions of the control system. As a result, the power of the reactor dropped to below 30 megawatts (thermal), and the reactor began to be poisoned by products of radioactive decay. That was the beginning of the end.

At this point I should say a few words about Anatoly Stepanovich Dyatlov, the deputy chief engineer in charge of operations for reactor units 3 and 4. Tall, lean, with a small angular face, gray hair smoothly brushed back, and sunken, dull, evasive eyes, Dyatlov joined the plant toward the middle of 1973. Bryukhanov gave me his curriculum vitae to study, before sending him along for a talk with me. It showed that he had been in charge of a physics laboratory at an enterprise in the Soviet Far East, where he appeared to have been working on small marine reactors. He confirmed this when we spoke.

"I studied the physical characteristics of the cores of small reactors," he told me.

He had never worked at a nuclear power station and was not familiar with the thermal layout of the plant or with reactors using uranium as a fuel and graphite as a moderator.

"How are you going to work?" I asked him. "This is a new facility for you."

"We'll learn," he replied in a strained voice. "Gate valves, tubing. It's simpler than the physics of the reactor."

Dyatlov had a curious manner. With his head leaning forward, his furtive dark gray eyes and staccato, tense voice, he seemed to squeeze out his words, between long pauses. It was not easy listening to him, and he seemed like a difficult person.

I reported to Bryukhanov that Dyatlov was not acceptable for the post of chief of the reactor section. He would encounter difficulties as manager of the operators not only on account of his personality and his obvious shortcomings in the art of communication, but also of his previous work experience, since he was a pure physicist with no knowledge of nuclear technology.

After listening to me in silence, Bryukhanov said he would think about it. A day later, Dyatlov was appointed deputy chief of the reactor unit. At some point or other, Bryukhanov had heeded my advice, by appointing him to a slightly lower position. Yet he was still in the reactor unit. In this, Bryukhanov committed a mistake—one events proved fateful.

My predictions about Dyatlov turned out to be correct: he was slow-witted, quarrelsome, and difficult. While I was working at Chernobyl, Dyatlov was not promoted. Moreover, I intended to have him moved in due course to the physics laboratory, which would have been a more suitable environment for him.

After I left, Bryukhanov began to promote Dyatlov, first to chief of the reactor section and later to deputy chief engineer for operations of the second phase of the plant.

Here are some comments on Dyatlov, from his subordinates who worked side by side with him for many years:

RAZIM ILGAMOVICH DAVLETBAYEV, deputy chief of the turbine unit, No. 4 reactor unit:

Dyatlov was a complex person, with a difficult character. Unlike most of the senior people at the plant, he kept to himself. He did not work particularly hard. In actual fact the section superintendents and their deputies were in technical control of the unit. If there was any need to resolve issues involving the participation of several subunits, they were resolved through horizontal contacts. Dyatlov did not mind, but we did. But there was no other way, as he constantly avoided difficult issues; even the start-up and running-in period of No. 4 reactor took place without his assistance and with no real guidance from him. While masquerading as a stern, demanding boss, in fact Dyatlov really did not care what was happening. The operators did not respect him. He rejected all proposals and objections that required any effort on his part. He was not involved with the training of the operators. Instead, he insisted that the individual sections should train them. He simply kept a record of their numbers. He began taking part in examinations one and half years after the start-up of No. 4 reactor unit, although as chairman of the commission he should have been taking part in them even before the start-up. He strictly punished mistakes and disobedience by the staff, by shouting and making everyone very nervous at the central control panel and at technical meetings. He had to spend a long time pondering the substance of questions, although he was quite a capable engineer. He seems to have been familiar with reactor technology. His knowledge of the technology of other technical sections was limited. With him in charge, people worked without any feeling of satisfaction. When not at work, he was sociable, a pleasant person to talk with, not without his own rather special brand of humor. He was a stubborn, tedious man, who did not keep his word.

VIKTOR GRIGORYEVICH SMAGIN, shift foreman of No. 4 reactor unit:

Dyatlov was a difficult, slow person. He usually told his subordinates: "I do not punish people right away. I think about the behavior of a subordinate for at least twenty-four hours, and then, when there is no longer any aftertaste left in my mind, I make my decision."

Dyatlov brought most of the senior physicists from the Soviet Far East, where he had worked himself as head of a physics laboratory. Orlov and Sitnikov, both of whom were killed, were also from there. And many others were his former colleagues.

Dyatlov could be unjust, even underhanded. Before the start-up of the reactor unit, during the period of assembly and the initial start-up procedures, I had an opportunity to leave for further study. "There's nothing for you to learn," Dyatlov told me. "You already know everything. But there are two colleagues of yours who really need to do some studying; they know very little." To sum up, throughout the assembly and initial start-up procedures, we did most of the hard work; but when the time came for jobs and salaries to be handed out, the highest salaries were awarded to those who had studied. When I reminded Dyatlov of his promise, he said, "They studied, but you did not."

Before the explosion, the general rule at the Chernobyl nuclear power station, which Dyatlov unfailingly applied, was that life was made extremely tough for the operational staff on the shifts, while the daytime, nonoperational staff of the various sections were pampered and encouraged. There were usually more accidents in the turbine hall and fewer in the reactor department. That accounts for the less vigilant attitude toward the reactor, the idea being that it was more reliable, and safer.

The issue is, therefore, whether Dyatlov was capable of immediately making the only correct assessment of the situation, just as the accident was about to occur. I believe that he was not. Moreover, he lacked two attributes indispensable for a manager of nuclear plant operators: caution and a sense of danger. On the other hand, he had more than enough self-confidence and disregard for both the operators as well as the safety rules.

It was precisely these latter qualities that dominated Dyatlov's conduct when, with the local automatic control system (LAR) switched off, the senior reactor control engineer, Leonid Toptunov, was unable to keep the reactor power at 1,500 megawatts, and it fell to 30 megawatts (thermal).

Toptunov had made a grave mistake. At such a low power level, the reactor begins to be intensively poisoned by products of decay such as xenon and iodine. It becomes difficult or even impossible to restore the parameters. All of this meant that the inertial rundown experiment had to be halted. This was clearly understood by all the nuclear operators, including the senior reactor control engineer, Toptunov and the shift foreman, Akimov, and even by Dyatlov, the deputy chief engineer for operations.

The situation in the room housing the main control panel of No. 4 reactor unit was one of high drama. With uncharacteristic energy, Dyatlov ran around the operators' control panels, spewing forth a torrent of curses and foul language. His normally hoarse, quiet voice now took on an angry, steely tone. "You goddamn idiots, you haven't a clue! You've screwed everything up, you boobs! You're ruining the experiment! I can't believe what a bunch of assholes you are!"

His rage was understandable. The reactor was being poisoned by decay products. There were two options: increasing the power immediately, or waiting twenty-four hours for the poisons to dissipate. He should have waited. Dyatlov, you overlooked the fact that the core was becoming poisoned faster than you had thought. Stop now, and mankind might still be spared the Chernobyl disaster!

But he was unwilling to stop. Still cursing, he dashed frantically around the control room, thus wasting valuable minutes. He ordered an immediate increase in the power of the reactor.

Dyatlov continued to rant and rave.

Toptunov and Akimov pondered their next move, and with good reason. Before falling to such low levels, the reactor had been functioning at 1,500 megawatts, or 50 percent of its rated power. The operational reactivity reserve in this situation was 28 rods (28 rods were in place inside the core). It was still possible to restore the parameters. The safety rules prohibited an increase of power if it had fallen from 80 percent with the same operational reactivity reserve, because in such circumstances the reactor becomes poisoned more quickly. But these figures—50 percent and 80 percent—were painfully close. Time was running out, and poison was spreading in the reactor. Dyatlov went on swearing. Toptunov did nothing. He realized that he would hardly be able to rise to the previous power level of 50 percent; and that even if he did, the number of control rods inside the core would be sharply reduced. Consequently the reactor would have had to be shut down immediately. Therefore, Toptunov made the only correct decision.

"I'm not going to raise the power!" he said firmly. Akimov supported him. Both men explained their apprehensions to Dyatlov.

"You lying idiot!" Dyatlov turned on Toptunov, shouting. "If it falls from 80 percent, the rules allow you to increase power after twenty-four hours, but you fell from 50 percent! The rules don't say you can't do it. If you don't increase power, Tregub will."

This was psychological warfare (Yuri Tregub was the shift foreman, who had been relieved by Akimov and was still present, intending to witness the tests). Admittedly, there is no way of knowing whether he would have agreed to increase the power level. Dyatlov, however, had guessed right: Toptunov, intimidated by his boss's shouting, acted against his own professional instinct. He was an inexperienced young man, of course, only twenty-six years old. What an ill-fated moment in his life! He must have been calculating the odds: "The operational reactivity reserve is twenty-eight rods. In order to offset the poisoning, we're going to have to remove five to seven rods from the reserve group. I might cause a power surge, but if I don't do what I'm told, I'll be fired." (These remarks were made by Toptunov in the Pripyat medical center before he was sent to Moscow.)

Toptunov began to increase power, thereby signing a death sentence for himself and many other comrades. That symbolic sentence also carried the clearly visible signatures of Dyatlov and Fomin. Other discernible signatures include those of Bryukhanov and many other more highly placed comrades.

To be quite honest, however, we must admit that the death sentence was implicit, to some extent, in the very design of the RBMK reactor. All that was needed was a certain confluence of circumstances making the blast possible. And those circumstances did come together.

But I am anticipating. There was still time for second thoughts. But Toptunov continued to increase the power of the reactor. Not until 1 A.M. on 26 April 1986 was it possible to stabilize it at 200 megawatts (thermal). All that time the reactor continued to be poisoned by decay products; a further increase in power was impeded by the low operational reactivity reserve, by then far below the regulation level.

To clarify matters, I should point out that the operational reactivity reserve is defined by the number of absorber rods fully inserted into the core, in the area of high differential worth. For RBMK reactors the operational reactivity reserve is considered to be 30 rods. With the emergency power reduction (AZ) system functioning, the speed of insertion of negative reactivity is 1 β per second,* or enough to offset the positive reactivity coefficients under normal operating conditions. According to the Soviet government's report to the International Atomic Energy Agency, the operational reactivity reserve consisted of between 6 and 8 rods; according to the dying Toptunov—who looked at the printout of the Skala machine, the centralized control system for the RBMK reactor, 7 minutes before the blast—there were 18 rods.

In either case, the effectiveness of the emergency power reduction system was greatly reduced, so that the reactor became difficult to control.

This was because Toptunov, in trying to climb out of the "iodine

---

*For an explanation of the beta symbol, see page 60.—Ed.

well," extracted a number of rods from the reserves which should have been left intact.

Even so, it was decided that the tests should continue, although the reactor was already difficult to control. Clearly, the two people who were primarily responsible for the nuclear safety of the reactor and of the plant as a whole—the senior reactor control engineer, Toptunov, and the shift foreman, Akimov—must have been supremely confident. Admittedly, they did have misgivings, and they did try to stand up to Dyatlov at the time of the critical decision; nonetheless, the paramount fact, in those circumstances, was that they were firmly convinced that they would succeed. They expected the reactor to come to their rescue. As we have already seen, the inertia of the usual conformist thinking also played a role. After all, in the preceding thirty-five years there had been no major accidents at a nuclear power station. And in any case, nobody had ever been told about the ones that had occurred. The cover-up had been diligent and thorough. The staff lacked any negative feedback about past events. And the operators themselves were young and insufficiently experienced.

VIOLATIONS OF NUCLEAR SAFETY RULES

Toptunov and Akimov, who came on duty that night, as well as the operators on all the preceding shifts on 25 April 1986, failed to show the proper sense of responsibility and blithely proceeded to commit serious breaches of the nuclear safety regulations. They must have truly lost touch with the hazards all around them and forgotten that the most important parts of a nuclear power station are the reactor and its core. Their main desire was to finish the tests as quickly as possible. They also could not have been particularly devoted to their work; if they had been, they would have pondered every move thoughtfully, showing the vigilance expected of true professionals. Without such devotion, it is better not to become involved with controlling a device as dangerous as a nuclear reactor.

The breaches of established procedures for the preparation and conduct of tests, and the carelessness in controlling the reactor, suggest that the operators had only a superficial understanding of the peculiarities of the technological processes occurring in a nuclear reactor. It was obvious that not all of them were aware of the details of the design of the absorbent rods.

Twenty-four minutes, 58 seconds remained before the explosion.

Here is a list of the most serious violations, either written into the program or committed during the preparation and conduct of the tests:

- In an attempt to climb out of the "iodine well," the operational reactivity reserve was reduced to below the permissible level, thereby making useless the reactor's AZ, or emergency power reduction, system.
- The LAR system (local automatic control) was erroneously switched off, thus reducing the power of the reactor to below the level prescribed in the program; the reactor became difficult to control.
- All eight main circulation pumps were connected to the reactor, with discharge levels on some of them at excessively high emergency levels; as a result, the coolant was close to saturation temperature (this was done in keeping with requirements of the program).
- For the purpose of repeating the experiment without electrical current, the reactor protection systems relying on the shutdown signal from two turbogenerators were neutralized.
- The protection systems triggered by preset water levels and steam pressure in the drum-separators were blocked, in an attempt to proceed with the test despite the unstable condition of the reactor; the reactor protection system based on heat parameters was cut off.
- The MPA protection system, for the maximum design-basis accident, was switched off, in an attempt to avoid spurious triggering of the ECCS during the test, thereby making it impossible to limit the scope of the probable accident.
- Both emergency diesel-generators were blocked, together with the operating and start-up/standby transformers, thus disconnecting the unit from the grid; though intended to ensure a "pure

experiment," this actually completed the chain of conditions necessary for an extreme nuclear disaster.

All of these factors appear in an even more sinister light when one considers the various unfavorable neutron-physical coefficients of the RBMK reactor, which has a positive reactivity void coefficient of 2 β (2 beta) and a positive reactivity temperature coefficient, as well as a defect in the design of the absorber rods of the reactor protection and control system.

In this reactor core, which was 23 feet (7 m) tall, the absorbent part of the rods was 16.4 feet (5 m) long, with hollow segments each 3.28 feet (1 m) long above and below the absorbent part. The lower tip of the absorber rod—which, when fully inserted, descended below the core—was filled with graphite. Such a design meant that when the control rods were lowered into the core, the graphite tip entered the core first, followed by the hollow segment 3.28 feet (1 m) long, and only then by the absorbent part. No. 4 reactor at Chernobyl had a total of 211 control rods. The Soviet government report to the International Atomic Energy Agency said that 205 rods were in the fully withdrawn position above the core; whereas the senior reactor control engineer, Toptunov, contended that 193 rods were above the core. The simultaneous insertion of so many rods into the core immediately causes a surge of positive reactivity, because before the neutron-absorbing sections of the rods enter the core, they are preceded by the graphite tips and the hollow segments, which displace the water in the SUZ (protection and control) channels, while doing nothing themselves to dampen the chain reaction.

The surge of reactivity in such conditions reaches 0.5 β, and in a stable controllable reactor is not very serious. However, when combined with various unfavorable factors, this addition can be fatal, as it may lead to an uncontrolled power surge.

The question is: Did the operators know about this, or were they in a state of blissful ignorance? I believe they knew, to some extent. At any rate, they should have known—especially Toptunov, the senior reactor control engineer. However, he was still young, and had not yet fully absorbed his theoretical learning. Since the shift foreman, Akimov, had never worked as a senior reactor control engineer, it is conceivable that he might not have known. But he had studied the reactor design and had passed examinations in order to get his job. Moreover, all of the operators might have failed to notice this fine detail of the control rod design, as it was in no way associated with lethal risk. Yet all along, the death and horror of the Chernobyl nuclear disaster lay hidden in precisely that design detail.

I also believe that Bryukhanov, Fomin, and Dyatlov—not to mention the designers of the reactor—did have a rough idea of the control rod design; but it never occurred to them that a future explosion lay hidden in certain segments at the tips of the rods, the most crucial part of the nuclear reactor's protection systems. It simply never occurred to anyone that death could come from a device that was supposed to protect.

Of course, reactors must be designed so that they shut down whenever unexpected power surges occur. This is the most sacrosanct rule in the design of any controlled nuclear device. The Novo-Voronezh VVER (pressurized water) reactor does fulfill that requirement.

Neither Bryukhanov nor Fomin or Dyatlov had grasped the possibility of this turn of events. Yet over the ten years in which the plant had been in operation, they would have had time to graduate from the physics and technical institute twice over and master every fine detail of nuclear physics. That, however, was for hardworking people passionately devoted to their subject, and not for those content to rest on their laurels.

The reader may find it useful to know that the only way to control nuclear reactors is by controlling the proportion of slow neutrons, designated by the Greek letter β (beta). The nuclear safety rules stipulate that the power of a nuclear reactor may be safely increased at a rate of 0.0065 effective β per second. Once the proportion of slow neutrons rises to 0.5 β, a prompt neutron power surge begins.

The operational staff's violations of the rules in respect to the reactor protection system, which I listed earlier, could release reactivity equal to at least 5 β, leading to the fatal explosive power surge.

Was this entire sequence understood by Bryukhanov, Fomin, Dyatlov, Akimov, and Toptunov? The first two probably did not understand. The last three must have known about it theoretically, but it meant nothing to them in practical terms, as can be seen from their irresponsible behavior.

As long as he could talk at all, before his death on 11 May 1986, Akimov kept repeating the same agonizing thought: "I did everything right. I don't understand why it happened."

All of this also suggests that training in emergency preparedness at nuclear power stations, and the theoretical and practical training of the staff, were being done very poorly—for the most part within a primitive management pattern which ignored the underlying processes in the reactor core at every moment during operations.

THE CHOICE OF REACTOR

How did such offhand, criminally negligent attitudes come about? Who programmed the possibility of nuclear disaster into the destiny of the Byelorussian–Ukrainian Woodlands? Why was a uranium-graphite reactor chosen for installation 80 miles from Kiev, the capital of the Ukraine?

Let me go back to October 1972, when I was working as deputy chief engineer at the Chernobyl plant. Many people were already asking this kind of question at the time.

One day in October 1972, Bryukhanov and I were on our way into Kiev, where we had been summoned by Aleksei Naumovich Makukhin, then Ukrainian minister of energy, who had nominated Bryukhanov for the post of director at the Chernobyl nuclear power station. Makukhin's training and professional background was in thermal power generation.

On the way into Kiev, Bryukhanov said, "If you don't mind, why don't we set aside an hour or two, so that you can lecture the minister and his deputies about nuclear power and the design of nuclear reactors? Try to put it in layman terms, because they, like me, don't understand much about nuclear power stations."

"I'd be delighted," I replied.

Makukhin was a commanding figure, whose oblong face, with its stony expression, was quite intimidating. He spoke in short bursts, in a manner that brooked no opposition.

I told my audience about the layout of the Chernobyl reactor, the component parts of a nuclear power station, and the distinctive features of that particular type of plant. I recall Makukhin asking, "Do you feel the right reactor has been chosen? I mean, Kiev is quite close."

"It seems to me," I answered, "that a VVER pressurized water reactor of the Novo-Voronezh type would have been much more suitable than a uranium-graphite reactor. A two-circuit plant is cleaner, with less extensive pipeline communications, and discharges are less radioactive. In other words, it is safer."

"Are you familiar with the arguments of Academician Nikolai A. Dollezhal? It is true that he does not advocate installing RBMK reactors in the European part of the country. But he makes a rather vague case. Have you read his conclusions?"

"Yes, I have. How shall I put it? Dollezhal is right. They shouldn't be put there. Those reactors have been used a great deal in Siberia, where they came to be known for their notoriously high levels of radioactivity. That's a serious argument."

"Then why didn't Dollezhal press his idea more forcefully?" Makukhin inquired.

"I don't know, Aleksei Naumovich," I said with a shrug. "Obviously there were certain forces which were more powerful than Academician Dollezhal."

"And what about the design-basis discharges from the Chernobyl reactor?" Makukhin asked, now plainly worried.

"Up to four thousand curies per twenty-four hours."

"And Novo-Voronezh?"

"Up to one hundred curies per twenty-four hours. That's a substantial difference."

"But what about the academicians? The operation of this reactor was approved by the Council of Ministers. Anatoly Petrovich Aleksandrov praises it, as the safest and most economical reactor. You, Comrade Medvedev, are laying it on a bit thick. Anyway, we'll manage just fine—it can't be *that* complicated. It's up to the operational staff to organize things in such a way that our first Ukrainian reactor is cleaner and safer than the one at Novo-Voronezh."

In 1982, Makukhin was transferred to the headquarters of the Ministry of Energy, as first deputy minister for the operation of electric power stations and grids.

On 14 August 1986, following the Chernobyl disaster, the Party Control Committee of the Communist Party Central Committee severely reprimanded the first deputy minister for energy and electrification, A. N. Makukhin, for failing to take the necessary measures to enhance reliability at the Chernobyl nuclear power station. But he was not dismissed.

The truth is that even in 1972 it would have still been possible to switch to a VVER pressurized water reactor at Chernobyl, thereby making the events of April 1986 much less likely. And the word of the Ukrainian minister of energy would have carried considerable weight.

Another typical episode is worth mentioning. In December 1979—when I was working in Moscow, at Soyuzatomenergostroy, the Department of Nuclear Power Plant Construction—I made an inspection trip to Chernobyl, to see how construction of No. 3 reactor unit was proceeding.

Vladimir Mikhailovich Tsybulko, at the time first secretary of the Kiev local committee of the Ukrainian Communist Party, attended the meeting of nuclear construction specialists. For a long time he said nothing, carefully listening to the discussion, and then took the floor himself. His scorched face, which bore traces of scars, from burn injuries sustained in the tank corps during the war, was deeply flushed. He stared straight ahead, focusing on no one in particular, and spoke as one not accustomed to objections. Nonetheless, his tone did betray touches of fatherly concern and goodwill. As I listened, I found myself thinking how readily officials in the nuclear power industry tend to hold forth on highly complex issues they do not fully understand, how quick they are to issue recommendations and "manage" a process that is a total mystery to them.

"Comrades, look what a beautiful city you have here in Pripyat," said the first secretary of the Kiev local party committee, pausing frequently (until then the meeting had been discussing the progress of construction of No. 3 reactor unit and the prospects for the construction of the entire plant). "You have been talking about four reactor units. I say that's not enough! I would build eight, twelve, and even all twenty reactor units! Then the town will grow to one hundred thousand inhabitants. It will be a fairy tale, not a town. You have a fine, experienced team of nuclear builders and assemblers right here. Why start a new construction site somewhere else? Why not build right here?"

During one of his pauses, one of the designers intervened to point out that an excessive concentration of nuclear reactor cores in one place was fraught with grave consequences, as it reduced the nuclear security of the state both in a wartime attack on the plants and also in the case of an extreme nuclear accident.

This sensible reply went unnoticed, whereas the proposal of Comrade Tsybulko was greeted enthusiastically as enlightening advice.

Shortly afterward, a start was made on the construction of the third phase of the Chernobyl plant, consisting of reactor units 5 and 6, and on the design of the fourth. However, 26 April 1986 was not far off; and at a single stroke, the explosion of No. 4 reactor knocked 4 million kilowatts of installed capacity out of the single nationwide grid and halted the construction of No. 5 unit, whose start-up seemed feasible in 1986.

Let us imagine what would have happened had Tsybulko's dream come true. In that case, all twelve reactor units would have been removed from the grid for a long time; a town with a hundred thousand inhabitants would have become deserted; and the losses of the state would have run into at least 20 billion roubles, and not 8 billion.

Also, the unit that blew up, No. 4, had been designed by Gidroproyekt with a potentially explosive reinforced leaktight compartment and pressure suppression pool beneath the nuclear reactor. At the time, when I was chairman of an expert commission dealing with the project, I categorically objected to such an arrangement and emphatically pressed for the removal of the potentially explosive structure from beneath the reactor. Nonetheless, the opinion of the experts was not heeded. As it turned out in real life, the explosion occurred both in the reactor itself and in the reinforced leaktight compartment beneath it.*

---

*I gave a detailed account of the expert opinion on the project in my story "The Expert Opinion," published in the Soviet-Bulgarian periodical *Druzhba* 6 (1986).

3

26 APRIL 1986

AFTER RETURNING ON the evening of 25 April from my visit to the Krymskaya (Crimean) nuclear plant, I looked through all my notes and the minutes of the meetings, concentrating on the synopsis of the meeting of the bureau of the Crimean local committee of the Communist Party of 23 April 1986, in which I had taken part.

Before the meeting of the bureau of the committee, I had talked with the head of its industrial section, V. V. Kurashik, and the industry secretary, V. I. Pigarev. I remember being surprised at the time that both comrades asked me virtually the same question: Wasn't the construction of a nuclear power station in the Crimea, the country's prime area for resort sanatoriums, being handled rather hastily? Weren't there any other more suitable places in the Soviet Union?

"There are," I had replied. "There are many sites on low-grade land, in areas with few or no inhabitants, where nuclear power stations could be built."

"Then why are they doing it here? Who makes these decisions?"

"The minister of energy, Gosplan. And Elektrosetproyekt plans the layout of the grid, according to the energy requirements of various regions."

"Yes, but we have electric power lines stretching for thousands of miles from Siberia to the European part of the country. Surely—"

"Yes, that's so."

"So it's possible they might not build in the Crimea?"

"Yes, they might not."

"They must not," said Pigarev, with a sad smile. "But we will build." The committee secretary pulled himself together.

"Yes, we will."

"Today there's going to be a serious discussion of this in the bureau. Both builders and management have been rather slack; they have not been keeping up with the plan. This just can't go on." Pigarev looked at me questioningly. "Can you give me some idea of how things are really going at the construction site, so that I can sound more convincing at the meeting of the bureau of the regional committee?"

I analyzed the situation, and the secretary made a convincing address.

On the night of 25 April–26 1986, all those who were later to be blamed for the nuclear disaster in Chernobyl were sleeping calmly: Ministers Mayorets and Slavsky; the president of the National Academy of Sciences, A. P. Aleksandrov; Ye. V. Kulov, chairman of the Nuclear Safety Committee; even Bryukhanov, director of the Chernobyl plant; and Fomin, its chief engineer. Moscow was asleep, as was the entire half of the world then in darkness. At the same time, in the control room of No. 4 reactor unit at Chernobyl, truly historic events were taking place.

Akimov's shift, it should be recalled, came on duty at midnight, 1 hour and 25 minutes before the explosion. Many shift members did not survive till morning. Two of them were killed instantly.

THE LAST SEVENTEEN MINUTES
AND FOURTEEN SECONDS

At 1 A.M. on 26 April 1986, as a result of crude pressure from the deputy chief engineer, Dyatlov, the power of the reactor in No. 4 unit had been stabilized at 200 megawatts (thermal). The reactor continued to be poisoned by decay products; a further increase in power was impossible; the operational reactivity reserve was far below the level prescribed in the rules and, as I have pointed out, stood at 18 rods, according to Toptunov, the senior reactor control engineer. Those figures were provided by the Skala computer 7 minutes before the AZ (emergency power reduction) button was pressed.

At this time, of course, the reactor was out of control and in danger of exploding. Thus, pressing the AZ button at any of the moments remaining until the historic point X would have led to a fatal uncontrollable power surge. There was no way of influencing the reactivity.

Seventeen minutes and 40 seconds were now left before the explosion—a long time, practically an eternity. A historic eternity. Thought travels at the speed of light. Throughout those 17 minutes and 40 seconds one could think back over an entire lifetime, over the whole history of mankind. Unfortunately the workers at Chernobyl had time only to unleash the explosion.

At 1:03 A.M. and 1:07 A.M., in addition to the six functioning main circulation pumps, one more pump was turned on from each side. The intention clearly was that, after the end of the experiment, there would still be four pumps in the circulation circuit to ensure the core was being adequately cooled.

I should explain at this point that the hydraulic resistance of the core and of the forced circulation circuit varies directly with the power of the reactor. As the level of power in the reactor was low (only 200 megawatts [thermal]), the hydraulic resistance of the core was also low. But with all eight main circulation pumps now in operation, the total water flow through the reactor rose to nearly 16 million gallons (60,000 m³) an hour—a serious breach of operating regulations in a plant with a normal flow of 12 million gallons (45,000 m³). Working in such a regime, the pumps could fail; the pipes in the circuit could begin to vibrate as a result of cavitation (that is, when the water boils with strong hydraulic shocks).

The sharp increase in the water flow through the reactor led to reduced steam formation, to a fall in steam pressure in the drum-separators to which the steam-water mixture is delivered from the reactor, and to undesirable alterations of other parameters.

Toptunov, Akimov, and the senior engineer for No. 4 unit, Boris Stolyarchuk, tried to maintain the reactor parameters (steam pressure and water level in the drum-separators) manually, but without complete success. By then steam pressure in the drum-separators had fallen by 5 to 6 atmospheres, and the water level was below the emergency parameter. Yet Akimov, with Dyatlov's consent, ordered the blocking of the emergency protection signal for those parameters.

Could the disaster have been averted in this situation? The answer is yes. All they needed to do was categorically to scrap the experiment, switch on the emergency core cooling system, and start up the emergency diesel generators, thereby securing a reserve supply of electricity in case all power was lost. Operating manually, one step at a time, they should have lowered reactor power until the reactor was completely shut down, while taking great care not to press the AZ button, which would have been the equivalent of an explosion.

This chance was let slip. The reactivity of the reactor continued its steady decline.

At 1:22:30 (1½ minutes before the explosion), Toptunov saw, from a printout of the program for rapid appraisal of the reactivity reserve, that it stood at a level requiring the immediate shutdown of the reactor. In other words, it showed 18 rods instead of the statutory 28. For a while he hesitated; after all, computers occasionally lie. He nevertheless reported his findings to Akimov and Dyatlov.

It was still not too late to halt the experiment and, while the core was intact, to carefully lower reactor power by manual means. But that chance was let go, and the tests began. It is worth noting that all the operators, apart from Toptunov and Akimov, who both found the computer data disturbing, were quite relaxed and confident as they went about their work. Dyatlov was also relaxed as he walked about the control room, urging the staff on: "Another two or three minutes, and it will all be over! Get moving, boys!"

At 1:23:04, the senior turbine control engineer, Igor Kershenbaum, in response to Metlenko's announcement "Oscillograph on!" closed the throttle valves of No. 8 turbine, and the generator rotor began its rundown. At the same time, the MPA button was pressed. This meant that both turbines, No. 7 and No. 8, had been shut down. The emergency reactor protection system designed to be triggered by the shutdown of two turbines had been blocked, so that the experiment could be repeated if the first attempt failed. This was another departure from the test program, which did not provide for the blocking of the reactor protection system triggered by the shutdown of two turbines. Paradoxically, however, if the operators had acted properly and not neutralized this safety feature, the shutdown of the second turbine would have triggered the emergency power reduction system, and the blast would have hit us 1½ minutes sooner.

At that point, at 1:23:04 the main circulation pumps began to steam up, and water flow through the core began to decline. The coolant in the fuel channels of the reactor began to boil. This process was initially slow, and the test had been under way for some time before the power slowly started rising. Who knows, maybe power might have continued to increase smoothly? There is no way of knowing.

Toptunov was the first to notice the power increase and sound the alarm. "We've got to trigger the emergency power reduction system, Aleksandr Fyodorovich, we're having a power surge," he said to Akimov.

Akimov quickly looked at the computer printout. The process was slow, really slow. He hesitated. Admittedly, there was another signal, showing that there were 18 rods instead of 28, but—The shift foreman was experiencing conflicting emotions. He had genuinely not wanted to increase power after it had fallen to 30 megawatts. In fact, the very thought made him sick and weak at the knees. He had been unable to stand up to Dyatlov; he just didn't have it in him. Eventually he had conceded, with profound misgivings. Once he had conceded, he became more confident. He had increased the power of the reactor from below the minimum permissible level, while all the time awaiting some new justification for pressing the emergency protection button. Such a moment now seemed to have come.

It may be assumed that the suppression of the emergency power reduction system (AZ) had been connected to the MPA button, because when that button was pressed the AZ rods for some reason failed to descend.

That may explain why, at 1:23:40, Akimov pressed the AZ button, in an attempt to duplicate the emergency signal. This is, however, merely an assumption, for which there is still no documentary confirmation or eyewitness testimony.

"I'm activating the emergency power reduction system!" Akimov shouted, and reached for the red button.

At 1:23:40, he pressed the level-5 emergency power reduction button, sending a signal that lowered into the reactor core all the control rods then in the fully withdrawn position, as well as the emergency protection rods themselves. However, the first thing to enter the core was those fateful rod tips, which, as we have seen, caused a 0.5 β increase in reactivity. And they entered the reactor at the precise moment when a sharp jump in reactivity was being caused by the extensive steam formation which had already begun. The increase in core temperature produced the same effect. So three factors inimical to the reactor core all came together at the same time.

Those extremely unwelcome 0.5 β were truly the last straw, as far as the reactor was concerned. At this point, Akimov and Toptunov should certainly have waited a little longer before pressing the button. Now the emergency core cooling system—which had been turned off, chained, and sealed shut—would have proved extremely valuable: it should have been allowed to do its job, through the main circulation pumps, of feeding cold water to the intake line, halting the cavitation and steaming-up process, thereby reducing the steam formation and perhaps even the release of excess reactivity. If only they had been able to switch on the diesel generators and the working transformer, to supply power to the electric motors of the top-priority systems and equipment, but, alas, it was not be. Such a command was not issued before the emergency protection button was pressed.

The button was pressed, and a prompt neutron power surge began in the reactor.

The rods began to descend, but stopped almost immediately. After that, shocks could be felt coming from the central hall. Toptunov stood there, panic-stricken. Seeing that the absorber rods, which were supposed to travel a total of 23 feet (7 m), had covered only 6.5–8 feet (2–2.5 m), Akimov dashed to the operator's panel and switched off current to the sleeves of the servo-drives, so that the force of gravity could lower the rods into the core. That did not happen—probably because the reactor channels had been twisted out of shape and the rods had jammed.

Then the reactor was destroyed. A substantial part of the fuel, the reactor graphite, and other structures from inside the reactor were blasted upward. But on the dials of the central control panel of No. 4 unit, as on the famous clock in Hiroshima, the needles were to be frozen forever in an intermediate position, showing a depth of insertion of 6.5–8 feet (2–2.5 m), instead of the requisite 23 feet (7 m); and in that position, they were to be entombed.

It was now 1:23:40.

When the AZ-5 button (level-5 emergency power reduction) was pressed, the lights of the dials on the Selsyn system flared menacingly. At moments like these, even the most experienced and unflappable operators feel a numbing thrust of apprehension. Deep inside the core, the destruction of the reactor had already begun, but the explosion was still to come. Twenty seconds were left to moment *X*.

The following people were then present in the control room of No. 4 reactor unit: the shift foreman, Aleksandr Akimov; the senior reactor control engineer, Leonid Toptunov; the deputy chief engineer for operations, Anatoly Dyatlov; the senior unit control engineer, Boris Stolyarchuk; the senior turbine control engineer, Igor Kershenbaum; the deputy head of the turbine section of No. 4 unit,

Razim Davletbayev; the director of the Chernobyl start-up enter-prise, Pyotr Palamarchuk; the foreman from the previous shift, Yuri Tregub; the senior turbine engineer from the previous shift, Sergei Gazin; two trainees of the senior reactor control engineer from earlier shifts, Viktor Proskuryakov and Aleksandr Kudryavtsev; and also the chairman of Dontechenergo, Gennady Petrovich Metlenko and two of his assistants, who were in adjacent nonoperational compartments of the control room.

Metlenko and his group intended to record the electrical charac-teristics of the generator during the rundown. Metlenko himself was supposed to monitor the rate of the slowdown in the rotations of the generator rotor on a tachometer, in the control room. A strange fate befell this man, who had been virtually left in the dark. Metlenko, who understood nothing about nuclear reactors, found himself in charge of an experiment that led to an extreme nuclear disaster. He did not even know the people with whom he was to work on the ill-fated night. He later had this to say: "I did not know the opera-tors. I met them for the first time that night, when we were brought together by the experiment. I had been expecting the test for days. It could have been held during the previous shift. I had to take readings. During the explosions I had no idea what was happening. I remember the operators were astonished. Why did it happen?"

What went through the minds of Akimov and Toptunov, the operators in charge of nuclear technological processes, when the absorber rods jammed halfway into the core, and the first sinister shocks were felt from the direction of the central hall? It is hard to say, because both operators died an agonizing death from radiation without leaving any testimony on the matter.

It is, however, possible to imagine what they must have experi-enced. I am familiar with the feelings of operators at the beginning of an accident. I have been in their place a number of times when doing operational work at nuclear power stations.

In the first split second, you experience a feeling of numbness, of complete collapse within your chest and a cold wave of fright—the main reason being that you have been taken by surprise, and that, to start with, you have no idea what to do—while the needles of the automatic printer drums and the monitoring instruments are swing-ing in all directions, while you are frantically trying to keep track, and while the cause and precise pattern of the accident remain unclear. At the same time you find yourself thinking, in the back of your mind, about who is responsible for the accident and about its consequences. A moment later, your head becomes remarkably clear and your composure is restored. You then move quickly and precisely to pinpoint the source of the accident.

Toptunov, Dyatlov, Akimov, and Stolyarchuk panicked. Kershenbaum, Metlenko, and Davletbayev understood nothing about nuclear physics, but the alarm of the operators soon spread to them, too.

The absorber rods had stopped halfway; they had not descended even after Akimov, the shift foreman, had disconnected the power supply to the sleeves of the servo-drives. Loud banging noises could be heard coming from the direction of the central hall. The floor was shaking. But the explosion was yet to come.

THE LAST TWENTY SECONDS

The time was now 1:23:40. For the 20 seconds remaining before the explosion, we shall leave the control room of No. 4 reactor unit at the Chernobyl nuclear power station.

At that precise moment, the foreman in charge of the reactor section on Akimov's shift, Valery Ivanovich Perevozchenko, entered the central hall, at level +50 (164 feet above the floor of the reactor building, and some 45 feet above the floor of the central hall),* on the balcony near the fresh fuel transfer station, on his rounds. He looked at the loading machine, which was standing still near the far wall, and then at the door in the wall behind which Kurguz and Genrikh, the central hall operators, were in a small compartment. He then looked down at the floor of the central hall, checked the spent fuel storage pools, which were heavily loaded, and turned toward the reactor lid, known affectionately as the *pyatachok*, or "five-kopek piece."

Pyatachok is the name given to the upper biological shield of the RBMK reactor, a circle 49 feet (15 m) in diameter, consisting of two

---

*The levels were marked on the walls of the huge reactor building.—Ed.

thousand cubes. Each of them, weighing 770 pounds (350 kg), sits like a cap on top of a fuel channel containing a fuel bundle. The *pyatachok* is surrounded by a stainless-steel floor made up of insulating blocks covering the compartments of the steamwater pipelines from the reactor to the drum-separators.

Suddenly Perevozchenko shuddered. Strong and frequent shocks began, and the 770-pound (350-kg) cubes—known by the technical term "assembly eleven"—started to jump up and down on top of the channels, as if one thousand seven hundred people were tossing their hats in the air. The entire surface of the *pyatachok* came to life, rocking in a wild dance. The insulating panels around the reactor shook and became bent, indicating a mixture of detonating gases beneath them was already exploding.

His hands torn by friction and colliding painfully with the corners of the handrail, Perevozchenko rushed madly down the steep, almost vertical spiral stairs, to level +10 (32 feet), into the corridor connecting the main circulation pump compartments—a descent equivalent to falling down a 130-foot (40-m) well.

Panic-stricken, his heart pounding wildly, realizing that something terrible and irreparable was happening, he ran on legs weak with terror to the left, to the door leading to the de-aerator, where beyond a safety lock 65 feet (20 m) farther on, lay the 328-foot (100-m) corridor, in the middle of which was the entrance to No. 4 control room. He rushed there to report to Akimov on events in the central hall.

While Perevozchenko was sprinting into the connecting corridor, Valery Khodemchuk, a machinist, was at the far end of the main circulating pump room, monitoring the conduct of the pumps during the rundown. The pumps shuddered violently, and Khodemchuk was about to notify Akimov, when a thunderous blast was heard.

At level +24, 79 feet above the floor of the reactor building, in compartment 604, beneath the reactor's feedwater machinery unit, Vladimir Shashenok, an adjuster from the Chernobyl start-up enterprise, was on duty with his instruments. He was taking readings from instruments during the rundown and in touch by phone with the control panel and the Skala computer complex.

What was going on in the reactor? To understand this, we have to turn the clock back a little and follow the chain of actions taken by the operators.

At 1:23, the reactor's parameters were as close as they ever were to stable. One minute before that, Stolyarchuk had sharply reduced the flow of feedwater to the drum-separators, thereby, as could be expected, increasing the temperature of the water entering the reactor.

Once the regulating valve had been closed, and the No. 8 turbogenerator switched off, the rundown of the rotor began. Due to the reduced flow of steam from the drum-separators, steam pressure began to increase slightly, at a rate of 0.5 atmospheres per second. The overall flow through the reactor began to decline because all eight main circulation pumps were being powered by the inertial force of the turbogenerator. Valery Khodemchuk could see them shaking. There was not enough energy; the power of the pumps was declining in proportion to the drop in the speed of revolution of the generator, thus causing a drop in the supply of water to the reactor.

The rise in steam pressure, on the one hand, and the reduction of both the water flow through the reactor and the delivery of feedwater to the drum-separators, on the other, were competing factors determining the steam content in the core and, accordingly, the power of the reactor.

As we have seen, the reactivity void coefficient (from 2 to 4 β) is more substantial in uranium-graphite reactors than in other types. The effectiveness of the emergency protection systems was thereby greatly reduced. However, as a result of the sharp drop in the flow of cool water through the reactor, the overall positive reactivity in the core began to rise. In other words, the rise in temperature led, on the one hand, to a rise in steam formation and, on the other, to a rapid increase in the temperature and void coefficients. That prompted the use of the emergency power reduction system. But as we have also seen, pressing the AZ button caused additional reactivity of 0.5 β. Three seconds after the AZ was pressed, the reactor power exceeded 530 megawatts, and the power surge lasted well below 20 seconds.

The increase in reactor power had the following effects: the hydraulic resistance of the core sharply increased; the water flow fell even farther; intensive steam formation occurred; there was film boiling*; the nuclear fuel assemblies were destroyed; the coolant, which by now contained destroyed fuel particles, came violently to the boil; and pressure rose abruptly in the fuel channels, which began to fall apart.

During the massive increase in pressure within the reactor, the feedback valves of the main circulation pumps burst, and the flow of water through the core stopped altogether. Steam formation increased. Pressure jumped 15 atmospheres per second.

Perevozchenko, the shift foreman for the reactor section, observed the massive destruction of the fuel channels at 1:23:40.

Then, in the final 20 seconds before the explosion, while Perevozchenko was hurtling down the steps from his original position, at level + 50 (164 feet), to level +10 (33 feet), a steam-zirconium reaction and other chemical and exothermic reactions were occurring in the reactor, resulting in the formation of hydrogen and oxygen—a highly explosive mixture.

At the same time, the reactor's main relief valves had been triggered, releasing a powerful burst of steam. However, this discharge was brief, as the valves, unable to cope with the enormous pressure and flow rate, were destroyed.

While this was happening, the gigantic pressure ripped apart the lower water and upper steamwater communication lines, thus breaching the insulation between the top of the reactor and the central hall and the drum-separator compartments, and between its bottom part and the reinforced leaktight compartment which the designers had intended as a means of confining a worst-case nuclear accident. Yet no one had expected the kind of accident that actually happened; for that reason, in the circumstances, the reinforced leaktight compartment simply acted as an enormous reservoir in which detonating gas began to accumulate.

---

*"Film boiling" is a most undesirable condition which occurs when bubbles at the surface of the boiling water in the fuel channels are replaced by a film that prevents the heat of the core from being transferred to the water. A safety margin is required at all times.—Trans.

THE EXPLOSIONS

At 1:23:58, the concentration of hydrogen in the explosive mixture in the various compartments reached the stage of detonation; and, according to several eyewitnesses, two explosions occurred one after the other, while other witnesses said there were three or more. What it really amounted to was that the reactor and the reactor building of unit No. 4 were destroyed by a series of powerful explosions of the detonating gas.

The blasts shook the building precisely as the machinist Valery Khodemchuk was at the far end of the main circulation pump room, and as Perevozchenko was running along the de-aerator corridor toward No. 4 control room.

Flames, sparks, and chunks of burning material went flying into the air above No. 4 unit. These were red-hot pieces of nuclear fuel and graphite, some of which fell onto the roof of the turbine hall where they started fires, as the roof was coated with tar.

In order to understand the quantity of radioactive substances ejected by the explosion into the atmosphere and onto the grounds of the plant, we must consider the nature of the neutron field 1 minute and 28 seconds before the explosion.

At 1:22:30, the Skala computer printed out the actual power distribution fields and the positions of all the absorber control rods. (Since the computer scans for 7–10 minutes at a time, it may have shown the status of the equipment 10 minutes before the explosion.) The general picture of the neutron field at the time of the reading showed a bulging field, in the radial-azimuth direction (across the diameter of the core), and vertically, on average, a double-hump field, with higher neutron power distribution in the upper part of the core.

Thus, if the machine is to be believed, a sort of flattened sphere of high energy emission some 23 feet (7 m) across and 10 feet (3 m) high had been formed. The prompt-neutron power surge, film boiling, as well as the destruction, melting, and then evaporation of the nuclear fuel first occurred precisely in this part of the core, which weighed about 50 tons. And it was that same part of the core that was blasted high into the atmosphere by the explosion of the detonating gas, and borne by the wind in a northwesterly direction, across Byelorussia and the Baltic republics and beyond the boundaries of the Soviet Union.

The fact that the radioactive plume moved at an altitude of between 3,300 and 36,000 feet (1–11 km) was indirectly confirmed by the testimony of a service technician named Antonov at Sheremetyevo airport, who reported that arriving aircraft were being decontaminated for one week after the explosion at Chernobyl. Today's airliners, of course, fly at altitudes of up to 43,000 feet (13 km).

In this way, about 50 tons of nuclear fuel evaporated and were released by the explosion into the atmosphere as finely dispersed particles of uranium dioxide, highly radioactive radionuclides of iodine-131, plutonium-239, neptunium-139, cesium-137, strontium-90, and many other radioactive isotopes with a variety of half-lives. In addition, about 70 tons were ejected sideways from the periphery of the core, mingling with a pile of structural debris, onto the roof of the de-aerator, the roof of the turbine hall where it adjoined No. 4 unit, and also onto the grounds of the plant.

Part of the fuel landed on equipment, on the substation transformers, on the generator bus-bars, on the roof of the central hall of No. 3 reactor unit, and on the plant's ventilation stack.

I must emphasize that the radioactivity of the ejected fuel reached 15,000–20,000 roentgens per hour; and that a powerful radiation field, practically equal to the radioactivity of the ejected fuel (the radioactivity of the nuclear explosion) was immediately formed around the damaged reactor unit.* With increased distance from the pile of structural and mechanical debris formed by the explosion, the radioactivity declined in proportion to the square of that distance.

Another point to bear in mind is that the evaporated part of the fuel formed a large atmospheric reservoir of highly radioactive aerosols, which was particularly dense and powerful in the vicinity of the damaged unit and, indeed, throughout the whole plant. This reservoir, which was rapidly growing, spread out in all directions, and, carried away by shifting winds, took on the shape of an enormous, sinister flower head.

---

*For permissible human doses of radioactivity, see page 214.—Ed.

Some 50 tons of nuclear fuel and 800 tons of reactor graphite (from a graphite stack weighing a total of nearly 1,700 tons) remained in the reactor vault, where it formed a pit reminiscent of a volcanic crater. (The graphite still in the reactor burned up completely in the next few days.) Through the holes that had by then been formed, pulverized nuclear debris filtered down into the space beneath the reactor, reaching the foundations of the building, as the lower water communication lines had been smashed by the explosion.

I have dwelt on these details in order to depict the true extent of radioactive contamination in and around the reactor unit, so that the reader may be able to visualize the horrendous conditions under which the firefighters and operational staff had to work, while they still did not realize what had actually happened.

To grasp the true magnitude of the radioactive release, one should merely remember that the atomic bomb dropped on Hiroshima weighed almost 4.5 tons; in other words, the mass of the radioactive substances formed when it was detonated amounted to almost 4.5 tons.

However, the reactor of No. 4 unit at Chernobyl spewed into the atmosphere almost 50 tons of evaporated fuel, thus creating a colossal atmospheric reservoir of long-lived radionuclides: in other words, ten Hiroshima bombs, without the initial blast and firestorm effects, plus almost 70 tons of fuel and some 700 tons of radioactive reactor graphite which settled in the vicinity of the damaged unit.

On the basis of preliminary findings, we can say that radioactivity in the vicinity of the damaged reactor unit ranged from 1,000 to 20,000 roentgens per hour. Admittedly, there were remote and sheltered spots where levels were significantly lower.

This being so, what is one to make of the reassuring statements made by Deputy Chairman Shcherbina of the Council of Ministers of the USSR, Yuri A. Izrael, chairman of Goskomgidromet (USSR State Committee for Meteorology), and by his deputy, Yu. S. Sedunov, at the press conference on 6 May 1986 in Moscow, when they said that radioactivity in the vicinity of the damaged reactor unit was only 15 milliroentgens (0.015 roentgen) per hour? Inaccuracy on this scale is, to put it mildly, unforgivable.

In the town of Pripyat alone, radioactivity on the streets throughout the whole of 26 April and several days thereafter measured between 0.5 and 1 roentgen per hour at all points: timely and truthful information and organizational measures would have saved tens of thousands of people from high doses of radiation. Later, I shall give a more detailed analysis of local contamination and exposure of the population from Pripyat to Kiev and Chernigov, because only with such an analysis is it possible to understand the heroism of those who fought to eliminate the consequences of the disaster, or the responsibility of those incompetents whose inept leadership actually caused it.

First, however, let us retrace our steps. The sequence, number, and location of the explosions of the detonating gas that destroyed No. 4 unit reactor and building are of great interest.

After the destruction of the fuel channels and the rupture of the steamwater and water communication lines leading to them, steam saturated with evaporated fuel, together with the products of radiolysis and a steam-zirconium reaction (hydrogen and oxygen), entered the central hall, the right and left drum-separator compartment, and the reinforced watertight compartment beneath the reactor vault.

After the rupture of the lower water communication lines, which carried cooling water to the core, the nuclear reactor was left entirely without water. Unfortunately, as we shall see, the operators either failed to understand this or were unwilling to believe it—thus setting in motion a whole sequence of errors, severe radiation, and death, which could have been avoided.

The explosions first occurred in the fuel channels of the reactor, as they began to fall apart from the massive increase in pressure. A similar fate awaited the reactor's lower and upper communication lines. As we have seen, the speed of the pressure increase itself was explosive: growing by 15 atmospheres per second, it soon reached 250–300 atmospheres. Yet the fuel channels and the communication lines had been designed to withstand a maximum pressure of 150 atmospheres, the ideal working pressure in the reactor channels being 83 atmospheres.

After wrecking the channels and entering the reactor space, which had been designed for a pressure of 0.8 atmospheres, the steam caused the metallic structures to collapse. The pipe supposed to extract steam from the reactor vault had been designed to cope with the destruction of one or two fuel channels; now all of them were lost.

Here is a passage from the log kept by one of the firefighters in No. 6 clinic, in Moscow: "When the explosion occurred, I was near the dispatcher's office, at the duty officer's post. Suddenly I heard a strong burst of steam. We thought nothing of it, because steam was being let off practically all the time I was at work.* I was about to go and take a break, when there was an explosion. I rushed to the window. After the first explosion, there were others."

So there was "a strong burst of steam . . . an explosion . . . after the first explosion, there were others."

But how many explosions were there? According to the firefighter, there were at least three. Or more.

Where could these explosions have taken place? The noise from the strong burst of steam meant that the reactor's relief valves had been triggered; but they were promptly destroyed, as were the steamwater and water communication lines a moment later. And the pipelines of the circulation circuit in the reinforced leaktight compartment may have been destroyed at the same time. Therefore hydrogen and steam first entered the compartments of the steamwater communication lines; then came the first weak shocks of the exploding mixture of detonating gas, which had been noticed at 1:23:40 by Perevozchenko, the shift foreman for the reactor section.

A mixture of hydrogen and steam also entered the right and left compartments of the drum-separators, the central hall, the reinforced leaktight compartment. A hydrogen content of 4.2 percent is sufficient in the air of a confined space to trigger the explosive hydrolysis reaction, which yields nothing but ordinary water.

Therefore, the explosions must have been heard coming from the right and left of the downcomer shafts of the reinforced leaktight compartments, from the right and left of the drum-separator compartments, and in the steam distribution corridor beneath the reactor. That series of explosions destroyed the drum-separator compartments, as well as the drum-separators themselves, each weighing

---

*He was referring to the triggering of the relief valves during normal operations at a nuclear power station.

130 tons, tearing them from their attachments and from the pipe-lines. The explosions in the downcomer shafts destroyed the right and left main circulation pump rooms. Valery Khodemchuk lies buried in one these.

This must have been followed by the big explosion in the central hall, which blasted away the reinforced concrete slab, the 50-ton crane, and the 250-ton refueling machine, together with the over-head crane on which it was mounted.

By exposing the reactor, which was full of hydrogen, to the air, the explosion in the central hall served as a kind of detonator. Both explosions, in the central hall and in the reactor, may have occurred simultaneously. In any case, then came the final, most terrible explosion of the detonating gas in the core, which was destroyed by the internal collapse of the fuel channels, being partly melted down and partly vaporized. This last explosion threw up a vast amount of radioactivity and red-hot chunks of nuclear fuel, some of which fell on the roof of the turbine hall and the de-aerator, and set the roof on fire.

Here is a further note from the log kept by the firefighter in No. 6 clinic in Moscow: "I saw a black fireball which swirled up over the roof of the turbine hall, next to No. 4 unit."

Or another note: "In the central hall* there was something that looked like a glow or a light. But at that spot there was nothing—apart from the reactor lid, the *pyatachok*—that could have been burning. Between us we decided that the light was coming from the reactor."

As they observed this scene, the firefighters were already on the roofs of the de-aerator and the specialized chemical unit, at level +71 (233 feet), where they had climbed to get a better picture of the situation.

The upper biological shield, weighing 500 tons, was hurled into the air by the explosion in the reactor. It came crashing back down, at a slight angle, leaving the core exposed to the air on both right and left.

One of the firefighters went up to level +35.6 (117 feet) in the central hall and looked down into the reactor. The mouth of the

---

*At level +35.6 (117 feet), although the floor of the central hall had ceased to exist.

volcano was spewing forth about 30,000 roentgens per hour, as well as powerful neutron radiation. These young men did not fully understand the magnitude of the radiation hazards to which they were exposed, although they may have guessed. The fuel and graphite over which they walked for some time, on the roof of the turbine hall, were also emitting up to 20,000 roentgens per hour.

These truly heroic firefighters extinguished the flames and conquered the blaze. But they were burned, many of them fatally, by another, invisible flame—the flame of gamma and neutron radiation—of a kind that no water could possibly extinguish.

EYEWITNESSES

There were very few people who, like the firefighters, saw from close up the explosions and the beginning of the disaster. Their testimony was truly historic.

At the time of the blast, forty-six-year-old Daniil Terentyevich Miruzhenko, a watchman, was on duty in the hydroelectric assembly office (Gidroelektromontazh), 1,000 feet (300 m) from No. 4 unit. He ran to the window on hearing the first explosions. Just then came the final, most terrible explosion, a thunderclap as loud as the sonic boom of a jet fighter, and a flash of light which cast a glow into the office where he was standing. The walls shook; the window-panes shattered, and some were blown out; the ground quaked beneath his feet. The nuclear reactor had just exploded. A pillar of flame, sparks, and red-hot fragments of something or other shot up into the night sky. Bits of concrete and metal structures could be seen tumbling about in the air, above the flames.

"What's all that about?" thought the panic-stricken watchman, his heart pounding in his chest, and a tight, dry sensation all over his body, as if he had just lost twenty pounds.

Carried aloft by the wind, the great swirling black fireball began to rise into the sky.

Immediately after the main explosion, fire broke out on the roof of the turbine hall and the de-aerator. Molten tar could be seen falling from the roof.

"The whole place is on fire. What the hell is going on?" the watchman whispered, still shaken by the explosions and the trembling earth. The first fire trucks were now reaching No. 4 unit from the plant fire station, where the firefighters had witnessed the beginning of the disaster through the window. These were vehicles from the No. 2 fire patrol under Vladimir Pravik.

Miruzhenko rushed to the telephone and called the plant construction office, but there was no reply. The clock showed the time as 1:30 in the morning. The watchman then phoned the director of Gidroelektromontazh, Yu. N. Vypirailo, but he also was not in. He was probably out fishing. Miruzhenko then prepared to wait until morning. I shall tell what eventually happened to him in a moment.

At the same time, on the far side of the plant, closer to Pripyat and the Moscow-Khmelnitsky rail line, a quarter of a mile (400 m) from No. 4 unit, Irina Petrovna Tsechelskaya, an operator at the cement-mixing section of the construction equipment unit attached to the Chernobyl plant, also heard the explosions—four of them—but stayed on her shift until morning. Her cement-mixing section supplied the construction crews building No. 5 reactor unit, located 4,000 feet (1,200 m) from No. 4 unit. On the night of 25–26 April, there were about 270 people at work at this site. The background radiation there was 1 or 2 roentgens per hour, but the air, both there and in the whole area, was heavily laden with short- and long-lived radionuclides and graphite dust, all of which were both highly radioactive and being inhaled by all those people.

When she first heard the explosions, Tsechelskaya found herself thinking, "Someone has broken the sound barrier. Perhaps it's a burst boiler in the start-up and standby boiler room. Or maybe a hydrogen explosion in the pipe vents?"

The first things that occurred to her were all drawn from familiar experience. But the start-up and standby boiler room was still there, undergoing an overhaul. Outside the weather was warm.

There was no aircraft noise to be heard, as was usual after a sonic boom. A hundred yards away, in the direction of Pripyat, a freight train rumbled by, and everything was again quiet.

Then the crackling of the flames soaring above the roof of the turbine hall, next to No. 4 unit, became audible. The bituminous finish of the roof, ignited by a nuclear firebrand, was burning.

"They'll put that out!" she decided confidently, going on with her work.

In the cement-mixing section, where operator Tsechelskaya was on duty, the background radiation stood at 10 to 15 roentgens per hour.

The radiation situation was least favorable to the northwest of No. 4 unit, toward the Yanov railroad station and the overpass across the tracks, leading from Pripyat to the Chernobyl-Kiev highway. The radioactive cloud passed this way after the blast at the reactor. The Gidroelektromontazh depot, where Miruzhenko was watching the explosions and events on the roof of the turbine hall, also lay in the path of the same cloud. It passed over the young pine grove between the town and the zone around the plant, liberally sprinkling it with nuclear dust as it went. The following autumn, and for some time to come, it was a "brown forest," lethal for all living things. Eventually it was leveled by bulldozers and buried. There used to be a trail through that pine grove, which people fond of hiking would take to and from work. I used to walk along that path on my way to work.

The external background radiation in the vicinity of the hydro-electric assembly depot was about 30 roentgens per hour.

Later, I shall refer again to the woes of Irina Petrovna Tsechelskaya, and the letter she wrote from Lvov, on 10 July 1986, to the Minister of Energy, Mayorets.*

A number of other people saw No. 4 reactor explode that ill-fated night of 26 April 1986. They were the fishermen who, practically day and night, seemingly in shifts, precisely because they went there in their spare time after completing their work shift, caught fish at the point where the plant outflow leads into the cooling-pond. After emerging from the turbines and the heat exchangers, the water was always quite warm, and the fish used to bite readily. Apart from anything else it was spring, spawning time, and so the waters were teeming.

This fishing spot was about 1.25 miles from No. 4 unit; the background radiation there was about half a roentgen per hour.

*See page 185.—Trans.

On hearing the explosions and seeing the fire, most of the people out fishing stayed there until morning, while others, vaguely alarmed by these events, returned to Pripyat, with dry throats and smarting eyes. The booming that normally accompanied the opening of relief valves sounded just like an explosion, so people had grown accustomed to ignoring such loud noises. As for the fire, someone would doubtless extinguish it. It was really nothing! Hadn't there been fires at the Armyanskaya and Byeloyarsk nuclear power stations?

At the time of the explosion, there were two more fishermen sitting on the bank of the feeder channel, trying to catch young fish, 260 yards (240 m) directly across from the turbine hall. Serious fishermen dream about catching fry like this. And without fry as bait it's better not to try for perch. Here the fry, especially in spring, manage to come closer and closer to the reactor unit, straight for the pump station, where they mass in great numbers. One of the two fishermen, by the name of Pustovoit, had no particular occupation. The second, Protasov, was a maintenance man who had been brought in from Kharkov. He thought very highly of Chernobyl; it had such clean air and wonderful fishing. He had even thought of taking up residence there permanently—if it could be arranged. It was, after all, in the Kiev region, where residence permits were hard to come by; it would be no easy matter moving there. He was catching plenty of fish fry that night and was in a good mood. It was a warm, starry Ukrainian night. It was hard to believe that it was still April, as it felt more like July. No. 4 reactor unit, a handsome snowy-white building, lay straight ahead. The unexpected combination of magnificent, dazzling nuclear power and the tender young fish wriggling in the net was a most pleasant surprise.

First they heard two dull explosions within the unit, which sounded as if they had come from underground. The fishermen could feel the ground shake. Then came a powerful steam explosion; and only after that did the reactor explode, with a blinding flash of flame and a firework display consisting of fragments of red-hot fuel and graphite. Pieces of reinforced concrete and steel beams went cascading through the air, blasted in all directions.

The fishermen's figures were, unknown to them, illuminated by nuclear light. Thinking that something had burst inside the plant, perhaps a gasoline tank, the two men went on fishing, not suspecting that they, just like the fry they hoped to catch, had themselves been caught in the powerful trap of a nuclear disaster. They watched with some curiosity as events unfolded. They could see with their own eyes as Pravik and Kibenok deployed their teams of firefighters, who then climbed up nearly 100 feet and attacked the fire.

"See that? One of them has got up on top of V block, more than 200 feet up! He's taken his helmet off! Fantastic! He's a real hero! You can see how hot it is over there."

A few hours later, as dawn approached, the two fishermen, each of whom had received a dose of 400 roentgens, became severely nauseated and both felt extremely ill. They had a burning sensation inside their chests, their eyelids smarted, and their heads felt as if they had just been on a wild drinking spree. And nonstop vomiting left them utterly exhausted. By morning their skin had turned black, as if they had been roasting in the sun at Sochi, on the Black Sea coast, for a month. They now had a nuclear tan, but still had not the faintest idea what was happening to them.

At daybreak they noticed that the men who had been up on the roof seemed also extremely sluggish and disoriented. That made them feel slightly better, as they were clearly not the only ones. But what had hit them all of a sudden? What could it be?

They made their way somehow to the medical center, and eventually were sent to the Moscow clinic.

Much later, one of them tried to make light of it, saying: "If you're ignorant, being curious can only get you into trouble, especially if your sense of responsibility is atrophied."

In the summer of 1986, Pustovoit, the man with no particular occupation, appeared on the cover of a foreign magazine and became famous in Europe. Misfortune can befall any living creature, but nuclear misfortune is all the more profound in that it runs counter to life itself.

Even next morning, on 26 April, more and more fishermen arrived at the same spot. The fact that they did shows how ignorant and careless people can be, how they had come to take emergencies for granted throughout all the years when news of such events was suppressed, and when those responsible were never punished. We shall return to our fishermen later in the morning, when the sun rose in the nuclear sky.

Now, before returning to the control room of No. 4 unit, I shall quote another eyewitness, G.N. Petrov, formerly equipment manager at Yuzhatomenergomontazh, a regional firm that installs the hardware in nuclear power stations.

I left Minsk for Pripyat, via Mozyr, in my car on 25 April 1986. In Minsk, I had been accompanying my son who was leaving for a tour of duty with the army in the GDR. My youngest son, a student, was on a construction crew in the south of Byelorussia. He also tried to reach Pripyat on the evening of 26 April, but the roads were blocked and they wouldn't let him through.

I approached Pripyat around 2:30 A.M. from the northwest, from the direction of Shipelichi. From Yanov station I could already see the fire above No. 4 unit. The ventilation stack, with its horizontal red stripes, was clearly lit up by the flames. I remember how the flames were higher than the shaft, so that they must have been nearly 600 feet in the air. Instead of turning to go home, I decided to go closer to No. 4 unit to get a better look. I approached from the direction of the construction office and stopped about 100 yards from the end wall of the damaged reactor unit. By the light of the fire, I could see that the building was half destroyed: there was nothing left of the central hall or the separator compartments; the drum-separators, which had been knocked out of position, shone with a reddish light. It was a terrible sight. Then I looked at the pile of debris and the destroyed main circulation pump room. There were fire trucks next to the reactor unit. An ambulance, with flashing lights, drove into town.

Incidentally, at the point where Petrov had stopped his car, the background radiation ranged between 800 and 1,500 roentgens per hour, mainly from the graphite and fuel ejected by the explosion and from the radioactive cloud.

I stood there about a minute, feeling a strangely oppressive sense of alarm, of numbness. Everything I saw was imprinted on my memory for the rest of my life. Despite myself, I got more and more profoundly alarmed and afraid. I felt some invisible threat looming. The air smelled just the way it does after a massive bolt of lightning, a lingering smoke which made my eyes smart and my throat dry. I suppressed a cough. Even so, in order to get a better look, I lowered the windows; it was a warm spring night. I could clearly see flames on the roofs of the turbine hall and the de-aerator; and I could see the firefighters, occasionally obscured by flames and smoke, and the hoses, stretching all the way up from the fire trucks below and shaking. One of the firefighters climbed onto the roof of V block, at level +70 (230 feet); he must have been checking on the reactor and coordinating the work of his comrades on the roof of the turbine hall, 100 feet (30 m) below him. Now, some time later, I realize that he was the first person in the history of mankind to be exposed to that kind of danger. Even in Hiroshima there was no one who got that close to the nuclear explosion, as the bomb went off at an altitude of 2,300 feet (700 m). But at Chernobyl he was right next to the explosion. At his feet the crater of a nuclear volcano was emitting 30,000 roentgens an hour. But at the time I was not aware of that. I made a U-turn and drove home to No. 5 district in the town of Pripyat. My family was asleep when I got there at around three o'clock in the morning. They woke up and said they had heard explosions, but didn't know what was happening. A woman from next door, whose husband had already been at the damaged unit, soon came rushing in to tell us about the accident and suggested we drink a bottle of vodka to decontaminate ourselves. We did just that, joked about it for a bit, and then went to bed.

At this point we shall interrupt Petrov, and hear from him later, on the evening of 27 April 1986.

IN THE CONTROL ROOM

We shall now return to the control panel of No. 4 reactor unit, which we left 20 seconds before the explosion, after Aleksandr Akimov had pressed the AZ button, and the absorber rods failed to descend into the core, having become jammed not even half-way down.

It is well here to remind the reader that, at numerous press conferences and in documents submitted by the Soviet Union to the International Atomic Energy Agency, the world was told that immediately before the explosion the reactor had been reliably shut down and the control rods had been inserted into the core. Large numbers of journalists then proceeded to repeat that lie, or thoughtless claim, and did so with every appearance of intelligence and authority. We also heard this from the deputy chairman of the Council of Ministers of the USSR, Boris Yevdokimovich Shcherbina, who stated that with the destruction of the reactor there was a "loss of criticality"—a novel concept in nuclear physics.

However, as we have seen, the major violations of the operating rules had practically neutralized the effectiveness of the emergency protection systems. When the AZ button was pressed, only 9.5 feet (2.5 m) of the absorber rods actually entered the core, instead of the 23 feet (7 m) required by the regulations; and far from shutting down the reaction, they actually helped start a prompt neutron power surge. Not a word was said, at any of the press conferences, about this exceedingly grave mistake by the reactor designers, which eventually proved to be the main cause of the nuclear disaster. But something *should* have been said, because the RBMK reactor was the nuclear time bomb, whose explosion signaled the death throes of an entire historical period.

So the core was destroyed.

"Is the fuel remaining in the core capable of a nuclear reaction, of a new explosion?" was the question put by the secretary of the Communist Party Central Committee, V. I. Dolgikh, to the deputy energy minister, G. A. Shasharin, on the night of 27 April 1986.

Those were the facts, and they could not be ignored.

1:23:58. A few seconds before the blast. Those present in the control room of No. 4 unit were in the following positions:

Toptunov and Akimov were at the left part of the operators' reactor control panel. Near them were the foreman of the previous shift, Yuri Tregub, and two young trainees, who had only recently taken their exams to qualify as senior reactor control engineers. They had gone there that night to see their friend Lenya Toptunov in action and to learn. Their names: Aleksandr Kudryavtsev and Viktor Proskuryakov. The AZ emergency power reduction button had been pressed 20 seconds previously. The positions of the absorber rods were displayed on the Selsyn system indicators—rather like alarm-clock dials—which were mounted on the operator's control panel. After the AZ button was pressed, both the senior reactor control engineer and the shift foreman stared in amazement as the background lights on the dials lit up and glowed as if they were red hot. Akimov rushed to the switch that would disconnect the power supply to the servo-drives which moved the absorber rods up and down; he pressed it, but the rods failed to descend and jammed, for all eternity, in an intermediate position.

"I don't get it!" shouted Akimov, deeply worried by what had happened.

Toptunov, his face deathly pale, was bewildered and distraught. He pressed buttons to supply feedwater to the fuel channels and to shift the film boiling margin. The channel control panel indicated that flows were at zero: in other words, that there was no water at all in the reactor, and that the film boiling margin was probably being lost.

The rumbling coming from the direction of the central hall suggested that film boiling had in fact started and that the channels were exploding.

Akimov once again exclaimed. "I don't get it! What the hell is going on? We did everything right."

A tall pale figure, his smooth gray hair combed straight back, walked over to the left part of the operators' control panel, the part dealing with the reactor itself. It was Dyatlov, the deputy chief engineer. He looked unusually confused, but his face still bore his standard expression, as if to say, "We did everything right. . . . This can't be. . . . We did everything right."

Boris Stolyarchuk, the senior control engineer of the unit, was at desk P, in the center of the control panel, which controlled the feed and de-aeration mechanisms. He was performing switching operations on the plant's de-aerator feed lines and regulating the delivery of feedwater to the drum-separators. He, too, was bewildered, though he was quite convinced he had done the right thing. The pounding shocks, which could now be heard coming from the bowels of the reactor building, were extremely unnerving. He felt something ought to be done to stop this sinister rumbling, but he did not know what to do, as he failed to understand the nature of what was happening.

Next to T desk, controlling the turbines (at the right end of the operators' control panel), were the senior turbine engineer, Igor Kershenbaum, and Sergei Gazin, a member of the previous shift who had stayed to see how things went. Kershenbaum was in charge of switching off No. 8 turbogenerator and starting the rundown. He acted in accordance with the approved program and the instructions of the shift foreman, Akimov. He was quite sure he was acting correctly. However, once he noticed that Akimov, Toptunov, and Dyatlov were losing their nerve, he too felt alarmed; but he had work to do, and there was no time to get excited. Together with Metlenko, he used a tachometer to monitor the pace at which the turbine rotor was running down. Everything seemed to be proceeding normally. Right there, at the turbine control panel, the senior man was the deputy head of the turbine section of No. 4 unit, Razim Ilgamovich Davletbayev.

To the left, the reactor control panel showed quite clearly that there was no water, and thus that the film boiling margin had been exceeded.

"What the hell!" thought Akimov, bewildered and indignant at the same time. "But all eight main circulation pumps are working!"

He then glanced at the voltage amperometers. The needles were hovering around zero.

"They're broken down!" he thought, with a sinking feeling inside, but only for a moment. He pulled himself together: "We've got to get some water in there."

At that moment, thundering noises came from right, left, and below, followed by one truly colossal explosion, apparently coming from all directions, which made it seem everything was falling apart: a gigantic shock wave carrying a white milklike dust, with the overwhelming pressure of the superheated radioactive steam, burst into the control room of what used to be No. 4 unit. The walls and floor crumbled as if hit by an earthquake; debris came crashing down from the ceiling. The sound of breaking glass could be heard coming from the de-aerator corridor; the light went out, leaving only the three emergency lamps on the accumulator battery; short circuits crackled and flashed; and all the electrical connections, as well as the power and control cables, were destroyed.

Dyatlov, trying to make himself heard above the infernal din, hoarsely ordered, "This is an emergency! Cool the reactor immediately!" It sounded more like a moan of sheer terror than an order. There was a noise of steam under great pressure and of hot water pouring out of damaged pipes somewhere. A flourlike dust invaded mouths, noses, eyes, and ears; mouths went dry, and everyone's senses were completely numb. This wholly unexpected intrusion had, like a bolt of lightning, suddenly suppressed their feelings of pain, fear, guilt, and irreparable calamity. Gradually they overcame their shock—first recovering the fearlessness and courage that come from desperation. Yet for a long time, almost until the moment of death, some of them were still very much under the influence of the reassuring comforting lies, myths, and legends created by backward, half-crazed minds.

Various thoughts flashed through Dyatlov's panic-stricken mind: "There's been a gas explosion. Where? Probably in the emergency protection and control system tank."

For some time this version of events, devised by a severely shaken Dyatlov, was given serious consideration; it consoled his ebbing consciousness and his paralyzed, though occasionally convulsive will. It even found its way to Moscow; and until 29 April, people believed it, using it as the basis for numerous decisions, some of which had lethal consequences. Why? Because it was the line of least resistance, providing justification and salvation for guilty parties from bottom to top, particularly for those who had miraculously survived in the radioactive bowels of the explosion itself. They needed their strength, and found it to some extent in a calm conscience. An unbearable night still lay ahead, but they had at least managed to vanquish the night of death.

"What's going on? What's all this?" shouted Akimov, when the cloud of dust had just begun to clear, the rumbling had ceased and the gentler sounds of hissing radioactive steam and pouring water were all that remained of the stricken nuclear giant's death throes.

Aleksandr Akimov—a powerfully built thirty-three-year-old, with a broad face and pink cheeks, wearing glasses, his dark wavy hair now covered in radioactive dust—stood there wondering what to do next.

"Sabotage? Impossible! We did everything right."

Leonid Toptunov—a chubby twenty-six-year-old with a ruddy complexion and a thin mustache—had been out of college for only three years. He was now pale and utterly bewildered, as if he expected something bad to happen, but was not sure where.

Perevozchenko ran into the control room, bruised and covered in dust.

"Aleksandr Fyodorovich!" he shouted to Akimov, gasping for breath. "Up there," pointing to the central hall, "there's something terrible going on. The lid of the reactor is breaking up. The blocks on top of the fuel channels on 'assembly 11' are hopping up and down. And those explosions! Did you hear? What is it?"

After the volcanic, deafening roar of the explosions, the whole unit was now oppressively silent, apart from the chillingly unfamiliar sounds of hissing steam and running water. The air smelled rather like ozone, only with a more bitter edge, causing a tickling sensation in the throat. Stolyarchuk, white as a sheet, looked helplessly yet expectantly at Akimov and Dyatlov.

"Take it easy!" said Akimov. "We did everything right. Something strange has happened." Then, turning to Perevozchenko, he said, "Valera, you run upstairs and see what is going on."

At that point, the door leading from the turbine hall to the control room burst open, and in ran the senior turbine engineer, Vyacheslav Brazhnik, a look of extreme alarm on his blackened face.

"The turbine hall is on fire!" he screamed, and then, saying something else which nobody could understand, he shot back out, toward the fire and a massive dose of radiation.

He was followed by Davletbayev and the chief of the group from the Chernobyl start-up firm, Pyotr Palamarchuk, who had come out that night to take readings of the vibration features of No. 8 generator, together with some people sent by the Kharkov turbine factory. Akimov and Dyatlov dashed to the open door. What they saw was horrendous, unspeakable. Fires were raging at levels +12 (39 feet) and zero. There was a pile of rubble on top of No. 7 turbine, as the roof had caved in. Burning oil was squirting from broken pipes onto the linoleum. Smoke was rising from the pile of wreckage. Red-hot graphite blocks and fragments of fuel lay at various points on the yellow linoleum, which was burning with a yellow sooty flame. Smoke, fumes, black ash falling in flakes, hot oil being ejected from smashed pipes, the severely damaged roof about to collapse at any minute, the protective paneling hanging perilously over the edge of the turbine hall—and noise, the crackling of fires raging above. A powerful jet of radioactive boiling water was pouring from the smashed feed pump and hitting the wall of the condensation compartment. At level zero there was a bright purple light caused by the arc formed between the two ends of a broken high-voltage cable. An oil line at level zero was broken, and the oil was burning. A thick column of black radioactive graphite dust was falling from the wrecked roof of the turbine hall onto the No. 7 turbogenerator, spreading out at level +12 along the floor and then drifting onto the people and the hardware below.

Akimov dashed to the telephone: "01! Quick! Yes, yes! The turbine hall's on fire and the roof, too. Yes! They're on their way? Great! Get a move on!"

Lieutenant Pravik's fire patrol had already deployed its trucks outside the turbine hall walls, and gone into action.

Dyatlov rushed out of the main control room and, his boots noisily crushing the broken glass underfoot, into the standby control room, across the way, next to the stair and elevator well. He pressed the level-5 AZ button and the switch to turn off the current to the electric motors. But it was too late. Why? The reactor was already destroyed.

Anatoly Stepanovich Dyatlov, however, thought otherwise. The reactor was really still intact. What had happened was an explosion of the protection and control system tank in the central hall. Yes, the reactor was intact.

The windowpanes in the standby control room were all broken; the broken glass made a squeaking noise underfoot, and the air smelled strongly of ozone. Dyatlov leaned out the window. It was night, and the crackling sound of a fire could be heard coming from above. In the reddish glow cast by the flames, he could see a monstrous pile of wreckage, consisting of structural debris, girders, smashed bricks, and concrete. Thick black objects lay strewn around on the asphalt, but it never occurred to him that they could be pieces of graphite from the reactor. It had been the same in the turbine hall, where there were also red-hot fragments of graphite and fuel. But his mind was unable to grasp the true meaning of what he had seen.

Torn between an overwhelming desire to do something and a feeling of immense apathy and despair, he went back to the control room.

On entering the control room, Dyatlov listened. Pyotr Palamarchuk was trying in vain to get through to compartment 604, where one of his subordinates, Volodya Shashenok, was stationed with measuring instruments. The line was dead. By this time, Palamarchuk had managed to find his way around No. 8 turbogenerator, gone down to level zero, and found the two young men from Kharkov in the mobile laboratory which had been installed in a Mercedes van. He insisted they leave the turbine hall, though two of them had already been close to the pile of debris where they had received a lethal dose.

In the meantime, Akimov had been calling all the heads of the various departments and sections, asking for help. Electricians were needed right away; there was a fire in the turbine hall. Hydrogen had to be extracted from the generators, and power had to be restored to the top-priority equipment.

"The main circulation pumps are dead!" he shouted to the deputy chief of the electrical shop, Aleksandr Lelechenko. "I can't start a single pump! There's no water in the reactor! We need help, quick!"

Davletbayev phoned Akimov and Kershenbaum from the booth in the turbine hall: "Don't wait for the electricians to get here. Get that hydrogen out of No. 8 generator right away!"

There was no answer from the dosimetrist, as the line had been cut. Only the municipal phones were working. None of the operators could feel, within themselves, the effects of the radiation. How great a dose were they receiving? What were the background levels? There was no way of knowing, as there were no instruments in the control room, and no "petal" respirators or potassium iodide tablets either.* It would have been wonderful if everyone could have taken a tablet right then. Who knows?

The line to the dosimetry panel was also dead.

"Petro, nip over there and find out why there's no answer from Kolya Gorbachenko," said Akimov to Palamarchuk.

"I have to get to Shashenok, he's in trouble. There's no answer."

"Right, get Gorbachenko and both of you go over to Shashenok."

Akimov then turned to other matters. Bryukhanov and Fomin had to be told what was happening. There were so many things to do. There was no water in the reactor. The SUZ rods (absorber rods of the protection and control system) were stuck halfway down. He was beginning to lose control of himself and oppressed by a feeling of shame. He broke out in cold and hot flashes as his racing mind tried to convey to him the full gravity of what had happened. The realization of the enormous responsibility involved suddenly hit him like a ton of bricks. He had to do something, everyone was looking to him for action. The two trainees of the senior reactor control engineer, Proskuryakov and Kudryavtsev, were just standing there idly. The control rods were stuck; so why not try to lower them by hand from the central hall? What a great idea! Akimov's spirits soared.

"Proskuryakov, Kudryavtsev!" he said, in the tone more of an appeal than a command, although he was fully entitled to tell them what to do. After all, everyone in the control room at the moment of the accident was under his direct command. But he nonetheless asked them, "Listen, fellows, go on over to the central hall, real fast. The SUZ rods have to be lowered by hand. There's something wrong over there."

Proskuryakov and Kudryavtsev set off. They were good men, young and totally blameless. Yet off they went to their deaths.

The first person to understand the full horror of what had happened seems to have been Valery Perevozchenko. He had seen the beginning of the disaster, and was already quite convinced that

---

*The "petal" is a particular type of Soviet respirator. Potassium iodide is a chemical that when, ingested, readily enters the thyroid gland. If taken in sufficient quantity before exposure to radioactive iodine, it can prevent the thyroid from absorbing it. Radioactive iodine, or iodine-131, when absorbed into the thyroid, can cause both cancerous and noncancerous growths.—Ed.

nothing could be done, and that truly horrendous destruction had taken place. From what he had seen in the central hall, he knew that the reactor had simply ceased to exist. So he realized that lives had to be saved, particularly those of his subordinates. This was how he, as shift foreman of the reactor section, saw his responsibilities. His first act was to rush off to find Valery Khodemchuk.

Testimony of NIKOLAI FYODOROVICH GORBACHENKO:

At the time of the explosion and later, I was in the dosimetry panel room. The whole place shuddered several times with tremendous force. I thought to myself, "We're finished!" But I looked again and saw that I was still standing there. There was one other comrade with me at the dosimetry panel, my assistant, Pshenichnikov, quite a young fellow. I opened the door to the de-aerator corridor and saw clouds of white dust and steam coming my way. There was a characteristic steam smell in the air. Detonations were still occurring, as well as short circuits. The dials for No. 4 unit on the dosimetry panel went out immediately, so there were no readings at all. I have no idea what was going on in that unit or what the radiation scene was like over there. On the dials for No. 3 unit (we have a single combined panel for both units in each construction phase), the emergency signals started flashing. All instruments went off the scale. I pressed the control room switch, but power had been cut off.

I could not get through to Akimov. I used the municipal phone line to speak to the dosimetry shift foreman, Samoylenko, who was at the radiation safety control panel of units 1 and 2. He then passed the message along to the people in charge of radiation safety, Krasnozhon and Kaplun. I tried to check the radiation levels in the room where I was and in the corridor, the other side of the door. All I had with me was a radiometer capable of handling up to 1,000-microroentgens-per-second. It showed a reading off the scale. I also had another instrument with a scale up to 1,000 roentgens per hour, but as soon as I switched it on, it burned out on me, just like that. That's all I had. I then went to the control room and reported the situation to Akimov. Everywhere the readings on the 1,000-microroentgens-per-second instrument were off the scale. That meant that in some places the dose was around 4 roentgens per hour, in which case no one could work there for more than about 5 hours. Obviously, those were the conditions you find in an emergency. Akimov told me I should walk around the unit and check the dosimetric situation. I went up to level +27 (89 feet) via the stair-elevator well, but no farther. Everywhere I went the instrument was off the scale. Along came Pyotr Palamarchuk, and together we went to compartment 604, to find Volodya Shashenok.

At the same time in the turbine hall, at level zero, fires were burning in a number of places. The ceiling had fallen in, chunks of red-hot fuel and graphite had fallen onto the floor and all over the machinery, and oil from a pipe smashed by a piece of concrete from the ceiling was on fire. Even the gate valve on the intake line of the feed pump was smashed, squirting radioactive boiling water toward the condensation compartment. The turbine oil tank and the hydrogen in the generator could explode at any minute. Something had to be done.

RESCUE WORK

We shall now leave the turbine hall, where the operational staff, at tremendous risk to their own lives, performed miracles of bravery and prevented the fire from spreading to other reactor units. That was an extraordinary feat, as heroic as the performance of the fire-fighters.

Meanwhile, acting on orders from Akimov, Proskuryakov and Kudryavtsev, the two trainees of the senior reactor control engineer, ran into the de-aerator corridor, where they turned right, as they were accustomed to doing, toward the elevator in the unit housing the auxiliary systems of the reactor section, but they saw that the elevator well had been destroyed and that the elevator, mangled by some unknown force, was lying on its side in a pile of construction debris. Then they turned back toward the stairs. There was a sharp smell of ozone in the air, as after a thunderstorm, only much stronger. They sneezed. And they sensed another mysterious force all around them; but even so, they kept going up.

Perevozchenko, who had notified Akimov and Dyatlov that he was going off to find his subordinates, who may have been in the pile of construction debris, rushed into the corridor just after them. First, he ran to the broken windows and looked up. His whole body could now feel the pervasive radiation; there was a marked smell of extremely fresh air, the kind of air you have after a thunderstorm, only far stronger. The fires on the roof of the de-aerator and in the turbine hall cast red reflections in the darkness outside. Normally you never feel the air, unless there is a wind; but Perevozchenko could now feel the pressure of some kind of invisible rays which were going right through his body. He was seized with a gut feeling of panic and terror, but his concern for his colleagues was uppermost in his mind. He leaned out a bit farther and looked to the right. At that point he realized that the reactor unit had been destroyed: where the walls of the main circulation pump rooms had once stood he could now dimly make out, in the darkness, a huge pile of construction debris, pipework, and wrecked machinery. Wondering how things were higher up in the unit, he stared up, to discover that the drum-separator compartments were also gone. There must have been an explosion in the central hall. He could see numerous fires burning there.

"We have no protection systems left—nothing," he thought angrily. With every breath he filled his chest with radionuclides. His initial depression had passed, but his lungs were still burning.

Perevozchenko felt as if his chest and face and the whole of his inside was on fire. "What have we done?" Valery Ivanovich wondered to himself. "Our boys are dying. Two operators, Kurguz and Genrikh, are still in the central hall, where there was an explosion. Valera Khodemchuk is somewhere in the main circulation pump rooms. Volodya Shashenok is in the instrument room beneath the reactor feed machinery. Which way should I run? Who should I try to find first?"

The need to clarify the radiation situation was the most pressing of all. Perevozchenko ran, slipping on bits of broken glass, to the radiation safety control panel, to Gorbachenko.

The dosimetrist was pale but calm.

"What are the background readings, Kolya?" Perevozchenko inquired, his face now burning with a brown fire.

"Well—in the 1,000-microroentgens-per-second range, the instruments are off the scale. The panels for No. 4 unit are out." Gorbachenko smiled apologetically. "Let's assume that it's somewhere around 4 roentgens per hour, though it's probably much more."

"How come you don't even have any instruments?"

"Well, we did have a 1,000-roentgen instrument, but it burned out. The second one is locked up in the safe, and Krasnozhon has the key. The only thing is, that safe is in the huge pile of debris out there. I looked. There's no way you can get to it. You know the way people used to think—no one ever seriously thought about a real disaster. No one believed it was possible. I'm going with Palamarchuk to look for Shashenok. There's no answer from him in compartment 604."

Perevozchenko left the dosimetry panel and ran to the main circulation pump room, where Valery Khodemchuk had been before the explosion. He was the closest one.

Petya Palamarchuk, the laboratory chief of the Chernobyl start-up firm, was running from the control room toward the dosimetry panel. He and his subordinates took readings of the performance and parameters of the various systems during the rundown. It was now evident that Shashenok was failing to respond, and that he was in the most dangerous place, compartment 604, right inside the thick walls of the reactor unit, where the massive blast had just occurred. How was he? The room he was in was vitally important, as the impulse lines from the main technological systems all converged on the sensors at that point. What if the membranes had been broken? There was steam at 570°F (300°C) and superheated water. He wasn't answering calls. Constant buzzing noises were audible over the receiver: perhaps the receiver had been cut off. Five minutes before the explosion, there had been a very good connection through to him.

Palamarchuk and Gorbachenko were already running toward the stair-elevator well.

"I'm going to get Khodemchuk!" shouted Perevozchenko, seeing them dash from the de-aerator corridor beyond the thick walls of the destroyed reactor section. They were about to enter an area littered with nuclear fuel and reactor graphite.

Palamarchuk and Gorbachenko ran up the stairs to level +24 (79 feet). Perevozchenko ran along the short corridor at level +10 (33 feet) toward the destroyed main circulation pump room.

At that moment, the two young trainees, Kudryavtsev and Proskuryakov, were slowly making their way past piles of debris to level +36 (118 feet), where the reactor hall was located. They could hear the crackling of flames and the shouts of firefighters coming from the roof of the turbine hall, amplified by the hollow elevator well, and also from somewhere near by, probably the lid of the reactor. "That's on fire, too?" they wondered.

At level +36 (118 feet) everything had been destroyed. The trainees made their way through the rubble of wrecked construction materials to the large room of the ventilation center, formerly separated from the reactor hall by a solid wall. It was clear that the central hall had been blown up like a balloon, and then its top had been torn away; its walls were now twisted out of shape, with metal bars sticking out like massive lacerations. In some places the concrete had collapsed, leaving the lattice of steel reinforcing bars plainly visible inside. The two young men stood there for a while, severely shaken and having some difficulty recognizing what had once been familiar sights. Although they experienced burning sensations in the chest and a terrible tightness across their temples, and their eyelids were smarting as if doused with hydrochloric acid, they nonetheless were overcome by a feeling of unusual and inexplicable excitement.

Along a corridor on the 50–52 axis they went, slipping occasionally on broken glass, to the entrance to the central hall, which was closer to the upper partition wall along row R. The narrow corridor was littered with debris and broken glass. Overhead the fire cast a red glow into the night sky; the air was filled with smoke and an acrid, oppressive smell of burning. Yet above all that, they sensed another unknown force in the air, a force pulsating with an insidious, stifling heat. That powerful nuclear radiation was ionizing the air, which henceforth became a new and frightening environment, unsuited for human existence.

With neither respirators nor protective clothing, they went to the entrance to the central hall and, passing through three half-open doors, entered what had once been the reactor hall, now littered with twisted wreckage and smoldering fragments. They could see fire hoses dangling over by the reactor. Water was streaming from the pipes. But there was no one to be seen, as the firefighters, utterly exhausted and fast losing consciousness, had retreated several minutes earlier.

Proskuryakov and Kudryavtsev found themselves practically in the mouth of the reactor, in terms of the radiation to which they were exposed. But where was the reactor? Could that be it?

The circular slab of the upper biological shield, with fragments of slender steel tubes (the fuel channel integrity monitoring system) jutting out in all directions, lay at an angle on top of the reactor vault. The steel reinforcing bars of the devastated walls hung shapelessly on all sides. This meant that the massive circular lid had been blasted off and then changed direction before crashing down again on the reactor. A red and blue fire was burning from the mouth of the reactor, with a powerful updraft. The trainees were hit in the face with nuclear heat which carried radioactivity of 30,000 roentgens per hour. They found themselves shielding their faces with their hands, as if the sun were too bright. It was obvious that no absorber rods were left at all; they must have been in orbit around the earth somewhere. There was nothing to lower into the core, nothing at all.

Proskuryakov and Kudryavtsev stayed near the reactor about a minute more, carefully remembering what they had seen. That was long enough for them to receive a lethal dose of radiation. Both died in great agony in No. 6 clinic in Moscow.

They then returned by the same route—feeling severely depressed and panic-stricken, now that the sense of nuclear-induced excitement had passed—to level +10 (33 feet), where they entered the control room and reported to Akimov and Dyatlov. Their faces and arms were the brown color characteristic of a nuclear tan. As the doctors found at the medical center, the skin all over their bodies was the same color.

"There's nothing left of the central hall," said Proskuryakov. "It's all been destroyed by the blast. You can see the sky overhead. There's fire coming out of the reactor."

"You guys must have got it wrong," said Dyatlov slowly and

indistinctly. "There was something burning on the floor, and you thought it was the reactor. Obviously an explosion of detonating gas in the emergency tank of the protection and control system (SUZ) has blown off the roof. Remember, that tank is mounted at level +70 (230 feet) in the outer surface of the end wall of the central hall. Yes, that's it. Well, that's not surprising. The volume of the tank is about 30,000 gallons (110 m³), that's quite a lot. An explosion like that could destroy not just the roof but the whole unit. We've got to save the reactor, it's still intact—we've got to get water into the core."

A myth was thereby created, about the reactor still being intact, about the site of the explosion having been the emergency water tank of the SUZ system, and the consequent need to deliver water to the reactor.

This myth was relayed to Bryukhanov and Fomin, and on to Moscow. The great deal of unnecessary and even harmful work done on this assumption made the situation at the plant even worse and increased the number of deaths.

Proskuryakov and Kudryavtsev were sent to the medical center. Fifteen minutes later they were followed by Kurguz and Genrikh, operators from the reactor hall, who had been next to the reactor when the explosions occurred. They were sitting at their duty stations after reviewing the central hall, and waiting for Perevozchenko to give them instructions for the whole shift. About 4 minutes before the explosion, Oleg Genrikh told Anatoly Kurguz that he was tired and was going to take a nap. He then went into a small windowless room nearby, measuring about 65 square feet (6 m²). Genrikh closed the door and lay down on a cot.

Anatoly Kurguz sat at his desk and made entries in his log. Three open doors stood between him and the reactor hall. When the nuclear reactor blew up, highly radioactive steam with nuclear fuel surged into the room where Kurguz had been sitting. He dashed through this inferno to try to close the door. Closing it, he shouted to Genrikh, "Fire! Everything's on fire!"

Genrikh leapt up from his cot and dashed to open his door. He opened it a little, but beyond it was such an unbearable smell of burning that he gave up trying, and instinctively lay down on the linoleum, where it was a little cooler, shouting to Kurguz, "Tolya! Lie down! It's cooler on the floor!"

Kurguz went in with Genrikh, and both of them lay down on the floor.

"There at least we were able to breathe. I had a funny burning sensation in my lungs," Genrikh later recalled.

They waited about 3 minutes, by which time the heat had abated somewhat—as it should, since there was now no roof over their heads. Then they went out into the corridor along axes 50–52. The severely blistered skin on Kurguz's bleeding face and arms was hanging off in strips.

Instead of going toward the stair-elevator well, from which the trainees Proskuryakov and Kudryavtsev were shortly to emerge, they aimed for the "clean staircase," and went down to level + 10 (33 feet). If they had run into the trainees, they would doubtless have turned them back, thereby saving their lives. But as it happened, they missed each other.

On the way to the control room, at level + 12 (40 feet), Genrikh and Kurguz were joined by two operators from the gas circuit, Simekonov and Simonenko. By now, Kurguz was in very poor shape: he was bleeding, and the skin under his clothes was covered with blisters. No one could do much to help him, as the slightest contact caused him intense pain. One wonders how he managed to walk the rest of the way. Genrikh's burns were less serious, as he had been protected by the windowless room. Both of them, however, had received a dose of 600 roentgens.

They were already on the de-aerator corridor when Dyatlov ran out of the control room to meet them.

"Go to the infirmary immediately!"

That meant a walk of between 490 and 550 yards (450–500 m) along the de-aerator corridor, to the administrative building situated in No. 1 reactor unit.

"Think you're going to make it, Tolya?" his colleagues asked Kurguz.

"I don't know. I doubt it. It hurts all over. Everywhere."

It was just as well that they decided not to go, as the infirmary of construction phase 1 was closed. There were also on that occasion no orderlies in construction phase 2. Bryukhanov was so confident everything was safe! Here we see the mentality of the "stagnant period" in action.

A first-aid vehicle was summoned to the administrative wing of the building housing reactor units 3 and 4 (phase 2 of the plant). Kurguz and Genrikh went down to level zero, punched out a windowpane that had miraculously survived, and emerged from the building.

Several times Dyatlov ran over to the control room of No. 3 unit, where he ordered Bagdasarov to shut down the reactor. Bagdasarov requested permission from Bryukhanov and Fomin, who refused to grant it. The operators from the central hall of No. 3 unit told their supervisor that alarm lights and buzzers had been triggered. They were under the impression that radioactivity had risen sharply, but did not realize that the fuel and graphite hurled by the blast onto the roof of their central hall was emitting radiation through the concrete roof.

On returning from one of his sorties, Dyatlov ordered Akimov, "Get on the phone to the daytime staff of each section! Especially the electricians, Lelechenko. We've got to stop the hydrogen from the electrolysis room going into No. 8 generator. They are the only ones who can do that. Get moving! I'm going around the unit."

Dyatlov walked out of the control room.

A number of times, Davletbayev ran from the turbine hall to the control room to report on the situation. It was full of all kinds of people. Samoylenko, the dosimetrist, took readings from Davletbayev with his instrument: "Razim, you're so radioactive it doesn't show on the scale! Change your clothes immediately!" As it happened, the turbine hall safety kit was under lock and key. Brazhnik, a man of immense strength, was told to break in with a crowbar.

Akimov commanded the senior unit control engineer, Stolyarchuk, and the machinist Busygin to turn on the feedwater pumps, to supply water to the reactor.

"Aleksandr Fyodorovich!" Davletbayev shouted. "We have no voltage! We need electricians here fast, so that the transformer at level zero can be started up. I don't know how they're going to do it, as the cables are all severed. Short circuits are flashing all over the place. There's an ultraviolet light at level zero near the feedwater pumps. It could be a piece of fuel or an arc from a short circuit."

"Lelechenko will be here soon with his boys! They'll take care of it!"

Davletbayev again plunged into the inferno of the turbine hall. At zero level, Tormozin was hammering wooden plugs into holes in an oil pipe; in order to complete the job, he sat on the pipe and burned his buttocks. Davletbayev then rushed off to No. 7 turbine, which was beneath a pile of wreckage, but was unable to get through as there was oil all over the linoleum. The sprinkler system was turned on, enveloping the turbine in a watery mist. The oil pump was turned off from the control panel.

Near No. 7 turbine there was a phone booth from which the machinists often called the control room. No one knew about the fragment of nuclear fuel which had landed on No. 5 transformer, the other side of the window, just across from the phone booth. This was where Perchuk, Vershinin, Brazhnik, and Novik received their lethal doses.

Meanwhile, the representative of Dontechenergo, Gennady Petrovich Metlenko—a short, frail-looking man with a pointed nose and drawn features—was waiting in the control room, with nothing particular to do. He had been assigned to the plant that night to supervise the inertial rundown experiment. Eventually Akimov noticed him and said, "Be a good fellow, go into the turbine hall and give a hand turning the gate valves. There's no power anywhere. It takes at least four hours to open or close each of them manually. The diameters are huge."

Metlenko quickly went to the turbine hall. Tragedy had struck there at level zero, where a falling girder had smashed the turbine oil pipeline, bringing hot oil into contact with pieces of red-hot reactor fuel which ignited it. The machinist Vershinin extinguished the fire and hastened to help his comrades prevent the oil tank from exploding and to make sure nothing else caught fire. Brazhnik, Perchuk, and Tormozin put out a number of fires elsewhere. The whole place was littered with intensely radioactive fuel and reactor graphite which had fallen into the turbine hall through holes in the roof; the highly ionized air was full of molten tar from the remains of the roof, black nuclear ash from burning graphite, radiation, and in general the smell of burning. A falling fragment of a girder had crushed the flange on one of the emergency feedwater pumps, which had to be disconnected from the suction and pressure lines of the de-aerators. It would take at least 4 hours to turn the gate valves by hand. Another pump had to readied for work on the "reactor," though that, too, would entail 4 hours of manual labor. The radiation fields at level zero in the turbine hall were between 500 and 15,000 roentgens per hour. Metlenko was sent back to the control room.

"We'll manage! Don't get in the way!"

Davletbayev arranged for the electricians of Akimov's shift to replace the hydrogen in the generator with nitrogen, so as to avert an explosion. Emergency oil was transferred from the turbine oil tanks into the emergency tanks on top of the reactor unit. The oil tanks were flushed with water.

On that ill-fated night of 26 April, the turbine operators performed a remarkable feat. Had they not acted as they did, the whole of the turbine hall would have been engulfed in flame, the roof would have collapsed, and the fire might have jumped to other reactor units, perhaps destroying all the other four reactors, with scarcely imaginable consequences.

By 5 A.M., when Telyatnikov's firefighters entered the turbine hall, everything had already been done. The No. 2 emergency feedwater pump had also been prepared for operation and even turned on to supply water to a nonexistent reactor. Akimov and Dyatlov assumed that the water was going into the reactor. However, that was impossible, for the simple reason that all the lower water communication lines had been severed by the explosion, and the water from the No. 2 emergency feedwater pump was really being delivered to the compartment beneath the reactor, into which a large amount of pulverized nuclear fuel had filtered. The intensely radioactive water, now mixed with fuel, flowed to the lower levels of the de-aerator, flooding the underground cable housings and the switchgear, causing short circuits, and coming close to cutting off all power to the three reactor units still in operation. All the reactor units of the Chernobyl plant were interconnected by means of the de-aerator gallery along which the main cables were routed.

By 5 A.M., Davletbayev, Busygin, Korneyev, Brazhnik, Tormozin, Vershinin, Novik, and Perchuk were all vomiting repeatedly and generally in poor condition. They were sent to the medical center. Davletbayev, Busygin, Korneyev, and Tormozin, each of whom had received a dose of about 350 roentgens, were to survive. Brazhnik, Perchuk, Vershinin, and Novik, with doses of 1,000 or more rad, died agonizing deaths in Moscow.

We shall now return to the beginning of the accident, and accompany Valery Ivanovich Perevozchenko on his fatal walk. Though he had gone looking for Khodemchuk, he was anxious to save all his subordinates. This was a man who knew no fear—whose courage and sense of duty led him straight into an inferno.

About this time Palamarchuk and Gorbachenko were working their way over the rubble in the stair-elevator well toward level +24 (79 feet) where Volodya Shashenok had last been heard from in compartment 604.

"What's the matter with him? Let's hope he's alive at least," Palamarchuk said to himself.

The unit was fairly quiet now, after the series of massive explosions; through the gaps in the roof, the crackle of the flames and frantic shouts of the firefighters could be heard coming from the roof of the turbine hall, as well as the unnerving moan of the burning graphite in the reactor. Those were the background noises, with radioactive water making sounds in the foreground like a babbling brook or a rainstorm, as it poured out in various places, both high and low, together with the now rather tired hissing sound of the last escaping radioactive steam. The air was extremely unusual: thick with highly ionized gas, smelling strongly of ozone, it burned throat, lungs, and eyes, causing a wrenching cough.

The two men rushed along without respirators, in pitch blackness, finding their way with the pocket flashlights all operators carried.

Perevozchenko ran along the short connecting corridor at level +10 (33 feet) leading to the main circulation pump room, where Valera Khodemchuk had been working, and stopped in his tracks, horrified. The room was no longer there: overhead was the sky, illuminated by the flames over the turbine hall, and straight ahead was a tangled pile of construction rubble, mangled hardware, and twisted tubing. The debris contained large amounts of reactor graphite, which was emitting at least 10,000 roentgens per hour. The speechless Perevozchenko passed his flashlight over the scene of devastation and wondered exactly how he had come to be there. How could anyone be in such a place? But his determination to save Valera prevailed. He strained to catch the faint sound of a moan or a voice.

Genrikh and Kurguz were also still up there somewhere, near the site of the explosion. He would save them, too. He could hardly abandon his own subordinates.

However, the longer he stayed where he was, the less were his own chances of survival. The body of the reactor shift foreman was absorbing more and more roentgens, and his nuclear tan was steadily turning darker in the darkness of the night. Not just his hands and face, but his whole body, beneath his clothes. His insides were on fire.

"Va-le-ra!" Perevozchenko yelled at the top of his voice. "Valera! Answer me! I'm here! Don't be afraid! We'll save you!"

He rushed straight toward the pile of rubble, climbing over pieces of wrecked masonry and patiently looking for openings, burning his hands on the fragments of fuel and graphite he happened to grasp in the dark.

Not the slightest sound indicated the presence of a human being, but he kept on looking, his skin bruised and torn by protruding metal reinforcing bars and sharp edges of concrete blocks. Eventually he made his way into compartment 304, but found it empty.

"Valera was on duty on the far side, I remember now."

Perevozchenko once again crossed the piles of rubble and went looking for Khodemchuk at the other end—to no avail.

"Valera-a!" yelled Perevozchenko, throwing his arms up and shaking his fists. Tears of frustration and grief rolled down his cheeks, now blackened and swollen by radiation. "Valera, come on! Answer me!"

There was no response, just the reflections of the fire blazing in the night sky above the turbine hall, and the piercing screams of the firefighters, which sounded like the calls of wounded birds. Up on the roof, death was infiltrating their bodies as they fought courageously.

Exhausted from a surge of nuclear fatigue, Perevozchenko went back down over the pile of rubble, staggered to the stair-elevator well and from there climbed to the central hall, at level +36 (118 feet), imagining that Genrikh and Kurguz were dying there, in a fiery nuclear hell.

He was unaware that some minutes earlier Anatoly Kurguz and Oleg Genrikh, who had miraculously survived the explosion, had left this devastated area and, though exposed to intense radiation and scalded by radioactive steam, descended the "clean" stairs to level +10 (33 feet); from there they had been sent to the medical center.

Perevozchenko retraced the steps of the trainees, Kudryavtsev and Proskuryakov, first entering the operator's cubicle; then, finding they were not there, he moved on to the central hall, thus walking straight into an additional blast of radiation from the noisily burning reactor.

As an experienced physicist, Perevozchenko realized that what had once been the reactor was now a gigantic nuclear volcano which could not possibly be extinguished by means of water, as the lower water communication lines had been severed from the reactor by the explosion; he also knew that the lives of Akimov, Toptunov, and the men who were in the turbine hall starting up the feedwater pumps, hoping to supply water to the reactor, were being lost to no useful purpose. There was no way water could be delivered to the reactor. The right course of action now was to save lives by evacuating everybody from the building.

Struggling to overcome fainting spells and intense nausea, Perevozchenko staggered down to the control room, and said to Akimov, "Sasha, the reactor has been destroyed! You've got to get everybody out of the unit altogether!"

"The reactor is still intact! We're going to get water into it!" Akimov countered angrily. "We've done everything right. You'd better get over to the medical center, Valera, you're not well. But you just got it all wrong, I assure you. It's not the reactor that's on fire, but parts of the building, and they'll put all that out."

At that very moment, while Perevozchenko was looking for Khodemchuk, who lay buried under a pile of rubble, Pyotr Palamarchuk and the dosimetrist Nikolai Gorbachenko, after struggling past piles of twisted metal and masonry at level 24 (78 feet), were finding their way to the measuring and monitoring instrumentation room, where Vladimir Shashenok had been at the time of the explosion. They found their comrade among the wreckage of compartment 604, pinned down by a fallen girder and severely scalded by steam and hot water. Although examination at the medical center later revealed that he also had spinal injuries and several broken ribs, now their immediate concern was to rescue him.

In the room where Shashenok had been immediately before the explosion, when pressure in the circuit was rising at 15 atmospheres per second, the piping and sensors were torn apart, releasing radioactive steam and superheated water, something fell, causing him to lose consciousness. He had severe thermal and radiation burns all over his body. The two young men freed him from the debris; then, with Gorbachenko's help, Palamarchuk, trying hard to avoid causing Shashenok additional pain, loaded him onto his back. With great difficulty they managed to reach level + 10 (33 feet). Then they took it in turns to carry their injured comrade about 500 yards (450 m) along the de-aerator corridor, until they reached the infirmary in the administrative building of No. 1 reactor unit. As it happened, the infirmary was closed. They called for an ambulance; and in 10 minutes an orderly, Sasha Skachok, arrived, and Shashenok was taken to the infirmary. Then the pediatrician, Dr. Valentin Belokon, came in his own ambulance, and remained on duty until morning, when he himself was taken to the medical center.

RADIOACTIVITY

And there Palamarchuk and Gorbachenko, who by carrying their comrade had themselves been exposed to severe radiation, soon found themselves as well. By then Gorbachenko, besides touring the unit measuring the background gamma radiation, had also somehow inspected the turbine hall as well as the outside of No. 4 unit. However, he could have saved himself the trouble. With the instrument he was using, which measured up to 3.6 roentgens per hour, he could not possibly register the actual radiation fields, which were fantastically powerful. And that meant, of course, that he could not warn his colleagues.

At 2:30 A.M., Viktor Petrovich Bryukhanov, the director of the Chernobyl plant, arrived at the control room of No. 4 unit. He was ashen and looked bewildered, indeed slightly frantic.

"What happened?" he asked Akimov in subdued tones.

The radioactivity in the air in the control room was then about 3 to 5 roentgens per hour. Closer to the pile of debris it was much higher.

Akimov reported that in his opinion there had been a serious radiation accident, but that the reactor was still intact; that the fire in the turbine hall was in the process of being extinguished; that Major Telyatnikov's firefighters were tackling the fire on the roof; and that No. 2 emergency feedwater pump was being readied for action and would soon be functioning. All that was needed was for Lelechenko and his people to supply electricity. The transformer had been disconnected from the unit to protect against short circuits.

"You're saying that there was a serious radiation accident but that the reactor is still intact. What's the radioactivity in the unit now?"

"Gorbachenko's radiometer is showing 1,000 microroentgens per second."

"Well, that's not much," said Bryukhanov, sounding only slightly reassured.

"That's right," said Akimov nervously.

"Can I go ahead and tell Moscow that the reactor is intact?" asked Bryukhanov.

"Yes, go ahead," was Akimov's confident reply.

Bryukhanov went to his office in No. 1 administrative block. From there, at 3 A.M., he phoned Vladimir Vasilyevich Marin, the nuclear power director in the Communist Party Central Committee, at home.

By that time the civil defense chief of the Chernobyl plant, S. S. Vorobyov, arrived at the damaged unit with a radiometer capable of measuring up to 250 roentgens—a distinct improvement. He realized from readings he took on the de-aerator, in the turbine hall, and near the pile of construction debris that the situation was extremely grave, as the needle went off the scale in several places.

Vorobyov passed on his findings to Bryukhanov.

"There's something wrong with your instrument," said Bryukhanov. "Fields that high are just impossible. Do you realize what that means? Get that thing out of here, or toss it in the garbage!"

"There's nothing wrong with the instrument," replied Vorobyov.

At 4:30 A.M., the chief engineer, Fomin, at last arrived at the control room. It had been difficult to get in touch with him: he did not pick up the phone himself at home, his wife had muttered something incoherent. Someone suggested he might be out fishing. His whereabouts were surely known to someone or other.

"What's this all about?"

Akimov reported on the situation, emphasizing the sequence of technical moves made before the explosion.

"We did everything right, Nikolai Maksimovich. I have no complaints about the people on the shift. By the time the level-5 AZ button was pressed, the operational reactivity reserve was 18 protection control rods (SUZ). The destruction was caused by an explosion of the 29,000-gallon (110-m^3) emergency watertank in the central hall, at level +71 (233 feet).

"Is the reactor intact?" Fomin asked in his pleasant baritone voice.

"It certainly is!" Akimov replied with assurance.

"Feed water into the reactor immediately!"

"The emergency feedwater pump is supplying water from the de-aerators to the reactor at this very moment."

Fomin walked away, torn by conflicting emotions. He was scurrying about mentally like a hunted animal, overwhelmed by the hopelessness of the situation; yet suddenly his ironclad confidence returned, and he felt sure they would pull through.

But he did not pull through. He was the first to crack under the monstrous weight of responsibility only now making itself felt, crushing the pathetic creature hidden beneath layers of arrogance and official confidence.

Dyatlov, the deputy chief operational engineer, left the control room and went down the stair-elevator well to the outside with the dosimetrist, having first instructed Akimov to feed water into the reactor. Chunks of reactor graphite, nuclear fuel, and debris lay strewn all over the asphalt. The air was thick and pulsating—a telltale sign of the presence of ionizing, intensely radioactive plasma.

"How's the radioactivity?" Dyatlov asked the dosimetrist.

"I'm off the scale, Anatoly Stepanovich," he replied, coughing violently. "My throat's terribly dry. At 1,000 microroentgens per second it's off the scale."

"You idiot! How come you don't have any instruments? You're not taking this seriously!"

"Yes, but who thought we would have fields like this?" the dosimetrist replied with a touch of annoyance. "There's a radiometer in the safe that can take readings up to 10,000 roentgens, but it's locked away, and Krasnozhon has the key. The trouble is, there's no way of getting to the safe anyway. It's under a huge pile of rubble, I saw for myself. And there's a lot of radiation out there. I can feel it without any instrument."

"You boob! You raving halfwit! You keep your instrument in a safe! Idiot! That's just incredible! So measure it with your nose!"

"That's just what I've been doing, Anatoly Stepanovich," said the dosimetrist.

"It would be nice if you knew what you were doing! I can also measure, damn it!" Dyatlov ranted on. "But that's not my job. That's your job. Understand?"

By now they were near row T and the auxiliary system unit, where a huge pile of debris reached all the way up to the separator compartments.

"What's all this?" Dyatlov exclaimed. "What the hell did they do? We're all finished!"

The dosimetrist was clicking the range switch back and forth, muttering, "Off the scale . . . off the scale . . ."

"Why don't you just chuck that stupid thing away right now, you jerk! Let's go and take a look outside the turbine hall."

Despite the poor light, it was unmistakably obvious that the asphalt was littered with pieces of graphite and nuclear fuel, some of the graphite being the size of footballs. With real radiation levels of around 15,000 roentgens per hour, it is not surprising that the dosimetrist's radiometer was off the scale.

Dyatlov and the dosimetrist were slow to register the evidence of their own eyes. As they walked around the end of the turbine hall, they saw nineteen fire trucks next to the concrete wall of the intake basin. The roar and crackle of the blaze on the roof of the turbine hall was now clearly audible; the flames were higher than the ventilation stack.

The deputy chief operational engineer of No. 4 unit curiously managed to sustain two quite different ideas, or images, of what he had seen. He found himself saying, on the one hand, "The reactor is intact! Feed in water!" And on the other hand, "There's graphite and fuel on the ground. But where did they come from? I can't figure out where they came from. The radioactivity is fantastic, I can feel it in my bones."

"Right! That's all!" Dyatlov commanded. "Let's get out of here!"

They went back to the control room. Gorbachenko went to his dosimetry panel. The deputy head of the radiation safety service, Krasnozhon, was due at any moment.

The overall exposure they received was 400 rads. They began suffering from headaches, extreme fatigue, and nausea at 5 A.M.; their skin had turned the brownish hue characteristic of a nuclear tan.

Gorbachenko and Dyatlov walked to No. 1 administrative building and from there were taken in an ambulance to the medical center.

Testimony of ALFA FYODOROVNA MARTYNOVA, wife of Marin, the nuclear power secretary of the Communist Party Central Committee:

At 3 A.M. on 26 April, we received an intercity phone call at home. It was Bryukhanov calling Marin from Chernobyl. After he had hung up, Marin told me that there had been a terrible accident at Chernobyl, but that the reactor was intact. He dressed quickly and sent for a car. Before leaving, he phoned the senior members of the Central Committee, in particular Frolyshev, who in turn called Dolgikh. Dolgikh then called Gorbachev and the members of the Politburo. Then he went to the Central Committee. At 8 in the morning Marin called home and asked me to get his things together as he was going on a trip—soap, tooth powder, toothbrush, towel, and so on.

116

At 4 A.M. on 26 April 1986, Bryukhanov received the following command from Moscow: "Make sure the nuclear reactor is kept cool all the time."

THE MYTH OF THE INTACT REACTOR

Nikolai Gorbachenko had now been replaced at the dosimetry panel of No. 2 phase, serving reactor units 3 and 4, by the deputy chief of the plant's radiation safety service, Krasnozhon. When asked by the operators how long they should stay on the job, he replied, in a matter-of-fact way: "In the range 1,000 microroentgens per second the instrument is off the scale. You can work for 5 hours, on the basis of a dose of 25 bers." (Thus it is clear that the deputy chief of the plant's radiation safety service had also failed to identify the true intensity of the radiation.)

Akimov and Toptunov also went up to the reactor several times, to see how well the No. 2 emergency feedwater pump was delivering water. But the fire was still blazing away.

Akimov and Toptunov were already brown from nuclear tan, their insides already wrenched by nausea. Dyatlov, Davletbayev, and the people from the turbine hall were already in the medical center. Akimov had already been replaced by the unit shift foreman Vladimir Alekseyevich Babichev; but Akimov and Toptunov stayed where they were, thus dooming themselves to certain death. Although their courage and heroism were inspiring, what they were doing was all based on the false assumption that the reactor was still intact. They were utterly unwilling to believe that the reactor was no more; that, instead of entering the reactor, the water was in fact mixing with nuclear particles and then pouring into the underground compartments, flooding the cable housings and high-voltage switching gear, and posing the risk of a blackout in the three reactor units still functioning.

"Something's keeping the water out of the reactor," thought Akimov. "There must be gate valves closed somewhere along the line."

When he and Toptunov reached the feedwater complex at level

117

+24 (78 feet), they found it half wrecked. At the far end there were holes in the masonry, through which the sky could be seen; the water flooding the floor contained nuclear fuel which brought the level of radioactivity to 5,000 roentgens per hour. How long could anyone live and work in such powerful radiation fields? Not for long, certainly. But these men were in a state of extreme excitation, with wholly exceptional powers of concentration; every last ounce of energy was being mobilized in response to a belated admission of their own guilt and responsibility, and to a sense of their duty to others. Their strength just came out of nowhere, as they kept on working when they should, in theory, have been dead.

The air throughout and around No. 4 unit was dense and pulsating, full of radioactive ionizing gas, saturated with the entire range of long-lived radionuclides the destroyed reactor was spewing forth.

With the greatest difficulty Akimov and Toptunov managed to open the regulating valves on two branches of the feedwater line some of the way; afterward, they went up to level +27 (90 feet), where they succeeded in forcing open two 11.8-inch (300-mm) gate valves, in a small pipeline compartment knee-deep in water saturated with fuel. Two more gate valves still needed to be opened, one each on the right and left branches of the pipeline, but none of them had any strength left—not Akimov nor Toptunov, or their assistants Nekhayev, Orlov, and Uskov.

After the explosion, the turbine operators in the turbine hall, the firefighters on the roof, and the electricians, under the deputy chief of the electrical section, Aleksandr Grigoryevich Lelechenko, all showed true heroism and a spirit of self-sacrifice. They confined the fire to the interior and the roof of the turbine hall, thereby saving the entire plant from disaster.

Aleksandr Grigoryevich Lelechenko went three times to the electrolysis room to disconnect the flow of hydrogen to the emergency generators, thus sparing the younger electricians from spending more time than strictly necessary in the area of intense radioactivity. The fact that the electrolysis room was next to the pile of radioactive rubble, surrounded by fragments of fuel and reactor graphite, with radiation of between 5,000 and 15,000 roentgens per hour, suggests the heroism and high moral fiber of that fifty-year-old man, who deliberately shielded his younger comrades with his own body. And then, while knee-deep in radioactive water, he studied the condition of the electrical switching gear, in an attempt to supply current to the feedwater pumps. His total exposure was 2,500 rads—enough to kill five people. Yet once he had been given first aid—in the form of an intravenous infusion—Lelechenko rushed back to the unit and worked there for several more hours. He died a terrible, agonizing death in Kiev.

Nor is there any doubt about the heroism shown by the reactor shift foreman, Valery Ivanovich Perevozchenko, by the maintenance man Pyotr Palamarchuk, and by the dosimetrist Nikolai Gorbachenko, all of whom did their utmost to rescue their comrades.

As for the actions of Akimov, Dyatlov, and Toptunov, and those who helped them, their efforts, however fearless and self-sacrificing, nonetheless had the effect of making the disaster worse, as they were based on a mistaken assumption: that as the reactor was intact, it had to be cooled, and water had to be fed in; and that the damage had been done by an explosion of the protection and control system tank in the central hall. This notion calmed Bryukhanov and Fomin who, having reported it to Moscow as their understanding of the situation, were instructed by way of response to feed water into the reactor without delay, so as to lower its temperature. On the other hand, those instructions also proved most reassuring and, to some extent, clarified the situation, by making it seem that all would be well provided the reactor could be supplied with water. In this way, they greatly influenced the actions of Akimov, Toptunov, Dyatlov, Nekhayev, Orlov, Uskov and others, who did their utmost to start the emergency feedwater pump and supply water to an imaginary "intact and undamaged reactor."

By offering a glimmer of hope, that same notion kept Bryukhanov and Fomin sane.

However, the reserve of water in the de-aerator tanks was running low, being down to only 127,000 gallons (480 m³). Admittedly, additional amounts were being fed in from the chemical purification system and from other reserve/standby tanks, thereby making it impossible to compensate for any losses of water from the three other working reactor units. Those other units, particularly No. 3 reactor, thus found themselves in an extremely difficult situation, which could well have degenerated to the point where the core could no longer be cooled.

In this connection, Yuri Eduardovich Bagdasarov, shift foreman at No. 3 unit, deserves credit for having both "petal" respirators and potassium iodide tablets on hand in his control room at the time of the accident. As soon as the radiation situation began to deteriorate, he ordered his staff to put on the respirators and take potassium iodide tablets.

Once he realized that all the water from the clean condensate tanks and from the chemical purification system had been switched to the damaged unit, he promptly notified Fomin that he was going to shut down the reactor. Fomin forbade him to do so. By morning Bagdasarov had himself shut down No. 3 unit and transferred the reactor to the cooling-off mode, supplying the water circulation circuit from the suppression pool. The action he took on his shift was courageous and of the highest professional standard, as it prevented the core of No. 3 reactor from melting.

Meanwhile, in the bunker of No. 1 administrative building, Bryukhanov and Fomin were constantly on the phone, Bryukhanov talking to Moscow and Fomin to the control room of No. 4 unit. The same version of events was sent out, over and over, to Marin, in the Communist Party Central Committee; to Minister Mayorets and to Veretennikov, the head of Soyuzatomenergo, in Moscow; and to V. F. Sklyarov, the Ukrainian minister of energy, and Revenko, the secretary of the Regional Committee in Kiev: "The reactor is intact. We are supplying water to it. There was an explosion in the emergency water tank in the central hall. The radiation situation is within normal limits. One man has been killed—Valery Khodemchuk. Vladimir Shashenok has suffered burns over 100 percent of his body and is in serious condition."

"The radiation situation is within normal limits." What can Bryukhanov have meant by this? Of course his instruments could not measure over 1,000 microroentgens per second, or 3.6 roentgens per hour. But what prevented him from having a sufficient number of instruments with a large range of measurements? Why were the crucial instruments locked up in the safe, and why were those used by the dosimetrist out of order? Why did Bryukhanov disregard the report of the plant's civil defense chief, S. S. Vorobyov, and fail to forward to Moscow his data for the radiation situation?

Cowardice, fear of being held responsible, and also, by virtue of his incompetence, refusal to believe that such a horrendous disaster could actually happen, all doubtless served to motivate Bryukhanov's actions. Yet the fact that the events at Chernobyl were totally beyond his understanding may explain, but does not justify, his conduct.

From Moscow, Bryukhanov learned that a government commission had been set up, and that the first group of experts would be leaving Moscow by air at 9 A.M.

"Hang in there! Cool the reactor!"

At times, Fomin began to break down, falling into a stupor one minute, ranting furiously the next, and crying, pounding the table with his fists and forehead, and rushing about in a frenzy of activity. His fine baritone now had a tense edge to it. He kept on badgering Akimov and Dyatlov, demanding an immediate flow of water to the reactor, and sent a succession of new people to No. 4 unit to replace those who were too sick to go on.

When Dyatlov was sent to the medical center, Fomin sent for Anatoly Andreyevich Sitnikov, the deputy chief operational engineer of construction phase 1, and said, "You're an experienced physicist. See if you can tell what state the reactor is in. You can be a neutral outsider, with no reason to lie. Please, go up onto the roof of V block and look down into the central hall, OK?"

Sitnikov went off to his death. He toured the entire reactor unit and went into the central hall, where he realized that the reactor had, in fact, been destroyed. However, feeling that this was insufficient, he climbed up to the roof of V block, the chemical water treatment plant unit, to get a better view of the reactor. The devastation below him was unspeakable. The massive reactor lid—covering much of the floor of the central hall—had been blasted off, and the pathetic remains of the battered concrete walls, their mangled reinforcing bars protruding wildly, looked like some monstrous flycatching plant, waiting for the chance to drag a living creature into its infernal belly. Dispelling this frightful image from his mind, but already feeling the embrace of hot nuclear tentacles on his face, hands, brain, and inner organs, Sitnikov carefully surveyed what was left of the central hall. The reactor had clearly been blown up. The upper biological shield, with jagged fragments of tubing, as well as clusters of severed cables, protruding from it in all directions, had obviously been hurled into the air by the explosion and come crashing down, at an angle, on the reactor vault. Through openings to right and left, he could see the blaze raging; the air was filled with an overpowering smell. The whole of Sitnikov's body, and particularly his head, was being riddled with neutrons and gamma rays. As he breathed the dense radionuclide gas, his chest began to feel as if it had been set on fire from within.

His head alone received as much as 1,500 roentgens. The radiation devastated his central nervous system. At the Moscow clinic he did not receive a bone marrow transplant, and, despite all measures taken to help him, he died.

At 10 A.M., Sitnikov notified Fomin and Bryukhanov that the reactor had, in his opinion, been destroyed. His report was angrily rejected and totally ignored: they went on pumping water into the "reactor."

FIGHTING THE FIRES

As we have seen, the first persons to be struck by the nuclear disaster inside the unit were the central hall operators Kurguz and Genrikh; the main circulation pump operator, Valery Khodemchuk; the maintenance man Vladimir Shashenok; the deputy chief of the turbine section, Razim Davletbayev; and the turbine machinists Brazhnik, Tormozin, Perchuk, Novik, and Vershinin. On the outside, the first ones who fearlessly tried to extinguish the fire were the members of Major Telyatnikov's team.

At the time of the explosion, the firefighter Ivan Mikhailovich was on duty at the plant fire station, 550 yards (500 m) from the damaged unit. When the alarm sounded immediately after the explosion, No. 2 fire patrol under Lieutenant Vladimir Pravik, who was in charge of fire protection at the plant, went to No. 4 unit.

Almost at the same time, No. 6 fire patrol under Lieutenant Viktor Kibenok, who was covering the town, left Pripyat for the plant. The fire brigade commander, Leonid Petrovich Telyatnikov, was on leave and was to return to work a day later. In fact, he was celebrating his birthday with his brother when a call came from the plant fire station.

"There's a fire in the turbine hall!" said the duty officer excitedly. "The signal came through from the plant. The roof's on fire. Lieutenant Pravik's crew has been sent over there. We've asked Pripyat to send Lieutenant Kibenok's crew to help out!"

"All right!" said Telyatnikov, approvingly. "Send a car around. I'll be right there."

He was driven there quickly. Once he had seen the fire, Telyatnikov realized that he did not have enough people on hand, and that help would be needed wherever he could find it. So he ordered Lieutenant Pravik to send a general alarm throughout the region. Over his two-way radio Pravik issued a No. 3-level alert, which required all fire trucks in the Kiev region, wherever they were, to proceed to the Chernobyl nuclear power station.

Shavrey and Petrovsky set up their trucks at row V and climbed their ladder to the roof of the turbine hall, where a fire and smoke storm was raging. Members of the No. 6 fire patrol, by now feeling thoroughly ill, were already there. The newcomers helped them to the ladder and then tackled the fire themselves.

V. A. Prishchepa set up his fire truck at row A and hooked up to the hydrant, and his crew climbed the ladder to the roof of the turbine hall. When they got there, they found that the roof was severely damaged, with some panels entirely missing and others hanging on by a thread. Prishchepa went back down to warn his comrades. On hearing the news, Telyatnikov issued orders for everyone to stay at their duty stations until victory was complete.

They did just that. Together with Shavrey and Petrovsky, Prishchepa stayed on the roof of the turbine hall until 5 A.M., after which they began to feel sick. In actual fact, they felt sick almost immediately but struggled on regardless, thinking that their condition was due to smoke and heat. When they descended at 5 A.M., however, they were feeling deathly ill; but by then the fire was out.

Another team of firefighters was also on the spot within 5 minutes of the accident, under the command of Andrei Polkovnikov, who went up to the roof twice to relay the instructions of Telyatnikov.

As Pravik had reached the scene of the disaster first, his entire team was sent into action on the roof of the turbine hall, while Kibenok's men, who arrived a little later, set about containing the fire in the reactor area. Here fires were burning at various levels, with five in the central hall alone. Kibenok, Vashchuk, Ignatenko, Titenok, and Tishchura fought these fires in the midst of a nuclear inferno. Once the blazes in the separator compartments and the reactor hall had been extinguished, there still remained one last fire—the most important of all—in the reactor itself. Initially, not realizing what they were up against, they poured water from hoses straight onto the raging nuclear reactor—naturally to no avail, as neutrons and gamma radiation are impervious to such treatment.

Pending the arrival of Telyatnikov, Lieutenant Pravik took charge of operations, personally checking on all details; this involved several trips to the reactor and also to the roof of V block at level +71 (233 feet) where Pravik was able to keep track of the rapidly changing situation.

The dense toxic smoke rising from the blazing tar of the roof sharply curtailed visibility and made breathing difficult. At any minute, sudden bursts of flame or structural collapse could have threatened the lives of the firefighters, who had to contend with intense heat, acrid air, molten tar which stuck to their boots, and black dust from radioactive graphite and keramzit, settling on their helmets. These extraordinarily brave men extinguished a total of thirty-seven fires in the reactor area and on the turbine hall roof.

Leonid Shavrey, from Pravik's detachment, stood at his post on the roof of V block, making sure that the fire did not spread. He felt immensely hot, both externally and internally. So far no one had given a thought to the possibility of radiation, on the assumption that this was a fire like any other, with no supernatural factors involved. Shavrey even took off his helmet. One by one, the men began to experience stifling pressure in the chest, severe coughing, nausea, vomiting, and fainting spells which made it impossible for them to continue. Around 3:30 A.M., Telyatnikov descended to the control room to report to Akimov on the situation on the roof. He wondered whether the extremely poor condition of his men might not be due to radiation, and asked for a dosimetrist. Gorbachenko told him that the radiation situation was complicated, and then instructed his assistant Pshenichnikov to accompany Telyatnikov.

The two men went up via the stair-elevator well, at the top of which there was a door leading to the roof; but it turned out to be locked. As they were unable to break it down, they then went all the way back down to level zero and across the street. Underfoot were fragments of graphite and fuel. Telyatnikov already felt sick, with nausea, vomiting, headache, and a severely discolored face, but attributed that to excess heat and the toxic smoke of the fire. Nonetheless, he wanted to be sure.

Pshenichnikov was carrying a 1,000-microroentgen-per-second radiometer which was off the scale everywhere he went, from ground level to the roof. With this instrument, whose maximum reading per hour was 3.6 roentgens, the dosimetrist could not have registered the actual levels of radiation, which at various points on the roof ranged from 2,000 to 15,000 roentgens per hour. The fact is that the roof had been ignited by falling chunks of red-hot fuel and graphite which, when mixed with molten tar, formed the viscous and intensely radioactive surface on which the firefighters had to walk.

As we have seen, things were no better on the ground, where graphite and fuel fragments were not the only source of radiation: there, nuclear dust from the plume formed by the explosion swept over everything.

Testimony of V. V. BULAV, a driver:

I had been told to report to Lieutenant Khmel. I did so, hooked up my truck to the hydrant, and turned on the water. My truck had just been overhauled and still smelled of fresh paint; the tires were also new. As I came close to the reactor unit, I heard something banging against the front right wheel. I hopped out to take a look. A piece of metal reinforcing bar had gone right through the tire and attached itself firmly to the rim. "What a damn shame!" I thought. I could hardly bring myself to look at it, as it had only just been overhauled. But while I was busy connecting the truck to the hy-

drant, there was no time to think about such things. Then I switched on the pumps and got back into the cabin, but I just couldn't get that chunk of metal out of my mind. As I sat there I could still see it sticking out of the tire, triumphantly. I decided I had to do something about it, so I got out of the truck and tried to yank it out, but it wouldn't come. I had to tug at it again and again. I ended up in the Moscow clinic with severe radiation burns on my hands. If I had known, I would have put gloves on. It's incredible.

The first firefighters to become incapacitated were Kibenok and his men, as well as Lieutenant Pravik.

By 5 A.M. the fire was out. The price, however, had been high, as seventeen firefighters, including Kibenok, Pravik, and Telyatnikov, were sent to the medical section and, that same evening, to Moscow.

A total of fifty fire trucks went to help at the site of the disaster, from Chernobyl and other regions in the Kiev region. Most of the work had already been done, however.

On that ill-fated and heroic night, the duty officer at the first-aid post of the Pripyat medical center was Dr. Valentin Belokon. He had been working in two teams with Aleksandr Skachok, an orderly. As Belokon was with a patient when the call came through from the stricken plant, Skachok went himself.

At 1:42 A.M., Skachok phoned from the plant, saying that there was a fire, that people had been burned, and a doctor was needed. Belokon left with Gumarov, a driver. Two other standby vehicles were also sent. On the way they passed Skachok, going in the other direction with emergency lights flashing; as it later turned out, he had been transporting Volodya Shashenok.

The door of the first-aid post in No. 1 administrative building had been nailed shut. They had to break it open. With pieces of graphite and radioactive fuel underfoot, Belokon went several times to No. 3 and No. 4 units. When Titenok, Ignatenko, Tishchura, and Vashchuk came down from the roof in a dreadful state, he gave them an injection of tranquilizer and sent them off to the medical center. The last ones to emerge from the flames were Pravik, Kibenok, and

Telyatnikov. By 6 A.M., Belokon himself was feeling sick and had to be taken to the medical center.

When he saw the firefighters, the first thing that struck him about them was their extreme agitation and nervousness. As he had never seen anything like it before, he gave them tranquilizers. It later became evident that they were suffering from a nuclear frenzy of the nervous system, a kind of spurious hypertonicity, which later changed to a deep depression.

Testimony of GENNADY ALEKSANDROVICH SHASHARIN, former deputy minister of energy and electrification of the USSR:

At the time of the explosion I was in Yalta, on holiday at a sanatorium with my wife. At 3 A.M. on the morning of 26 April, the phone rang in my room. It was the Yalta party office saying that a serious accident had occurred at the Chernobyl nuclear power station, that I had been appointed chairman of a government commission, and that I had to fly to Pripyat without delay to go to the scene of the accident.

I quickly dressed, went to the administrative officer on duty, and asked to be put in touch with the manager of Krymenergo in Simferopol and also with Soyuzatomenergo in Moscow. They put me in touch with Soyuzatomenergo, where I found G. A. Veretennikov already on the job, about 4 A.M. I asked him whether the emergency power reduction system had been triggered, and whether water was being supplied to the reactor. Veretennikov replied that both were being done.

Then the administrator of the sanatorium brought me a telex from Minister Mayorets, to the effect that Boris Yevdokimovich Shcherbina, deputy chairman of the USSR Council of Ministers had already been appointed chairman of the government commission, and that I too was to be in Pripyat on 26 April. I was to fly there right away.

I talked to the manager of Krymenergo, asked for a car at 7 A.M. and a reservation on the plane for Kiev. I spoke to Krymenergo from the Yalta party office, where the duty officer put me through.

I was picked up at 7 A.M. and taken in a Volga to Simferopol. I reached Simferopol just after ten. The flight for Kiev was not due to leave until eleven, so I had a little time and dropped in on the party regional committee. Nobody knew anything there; they just said they were worried about having nuclear power plants built in the Crimea.

I landed in Kiev about 1 P.M. The Ukrainian minister of energy, Sklyarov, told me that Mayorets with his team would be flying in any moment, and that I should wait. I had had only five days of my vacation.

Testimony of VIKTOR GRIGORYEVICH SMAGIN, shift foreman in No. 4 unit:

I was supposed to take over from Aleksandr Akimov at 8 A.M. on the morning of 26 April 1986. I had slept soundly during the night and heard no explosions. At 7 A.M. I woke up and went out onto the balcony for a smoke. From the fourteenth floor I had a good view of the plant. As soon as I looked that way, I realized that the central hall of my dear old No. 4 unit had been destroyed. Flames and smoke could be seen above the building. Clearly something terrible had happened. I rushed to the phone to call the control room, but the line was dead, probably to prevent information from leaking out. I got ready to leave. I ordered my wife to keep the doors and windows closed tight and the children indoors. She also was not to go out, but to wait home until I got back.

I ran to the bus stop, but no bus came. Soon a minibus came along, and I was told that I would be taken not to No. 2 administrative building, as usual, but to No. 1 administrative building.

When we reached No. 1 administrative building, we found it had been cordoned off, and the police were not letting people through. Then I showed my 24-hour pass as a member of the operational staff, and they grudgingly let me through.

Near No. 1 administrative building, I met V. I. Gundar and I. N. Tsarenko, deputies of Bryukhanov, who were on their way to the bunker. They told me to go at once to No. 4 control room and replace Babichev, who had taken over from Akimov at six that morning, and was probably exhausted by then. They reminded me

*to change in the "glass house." That was what we called the confer-
ence room. It occurred to me that if I had to go there to change,
there must have been radiation at No. 2 administrative building.*

*I went up to the "glass house." There was a mass of clothing such
as overalls, boots, "petal" respirators. As I was getting changed, I
could see the deputy minister for internal affairs of the Ukrainian
SSR, General G. V. Berdov, on his way into Bryukhanov's office.*

*I changed quickly, little suspecting that when I returned it would
be to the medical center with a severe nuclear tan and a dose of 280
rad. At that moment, rushing to put on my cotton overalls, cap,
boots, and petal-200 respirator, I was about to dash along the long
de-aerator corridor, which connects all the units, on my way to No.
4 control room. There was structural damage in the vicinity of the
Skala computer room, with water and steam leaking. There was
nobody in the Skala room; water—which I later discovered was
highly radioactive—was pouring from the ceiling onto the cabinets
containing the hardware. Yura Badayev must have been taken away.
Farther on I glanced into the dosimetry panel room, where Krasnozhon, deputy head of the radiation safety service, was in charge.
Gorbachenko was gone, perhaps on his rounds somewhere, though
he, too, may have been taken out of there. Krasnozhon was not
alone. He and Samoylenko, the night shift foreman of the dosimetrists, were having a furious argument over the fact that they could
not measure the radiation levels. Samoylenko was convinced that
they were enormously high, while Krasnozhon argued that it would
be possible to work in such an environment for 5 hours, on the basis
of a dose of 25 bers.*

*I interrupted their argument by asking how long it would be
possible for us to work. Krasnozhon assured me that the background
level was 1,000 microroentgens per second, or 3.6 roentgens per
hour, and that we could work for 5 hours on the basis of a 25-ber
dose. Samoylenko took the view that this was absurd, whereupon
Krasnozhon lost his temper again. I asked whether those were the
only radiometers they had. Krasnozhon told me that they had some
but they were in the safe, which had been buried in rubble after the
explosion. He felt that the people in charge of the plant had never
expected such a serious accident.*

"What about you? Aren't you in charge, too?" I thought as I moved on.

All the windows in the de-aerator corridor had been blown out by the explosion. There was as strong smell of ozone in the air, and I could feel the radiation going through my body, though they do say that we have no such sensory organs. Obviously, though, something was going on. I had an unpleasant sensation in my chest—a kind of spontaneous panic—but I did not let it get the better of me. It was now light outside, and the pile of construction debris could be seen clearly. The asphalt all around was littered with something black. I looked a bit closer and suddenly realized that it was reactor graphite! There was no mistaking it! I knew then that the reactor was utterly finished, but I still had not grasped the full gravity of what had actually happened.

I went into the control room, where I found Vladimir Nikolaye-vich Babichev and the deputy chief scientific engineer, Mikhail Alekseyevich Lyutov, who was sitting at the shift foreman's table.

I told Babichev that I had come to relieve him. It was 7:40 A.M. He replied that he had started his shift an hour and a half previously, and that he felt fine. In such instances, the new shift takes orders from the one already in place.

Babichev said that Akimov and Toptunov were still in the unit, opening gate valves in compartment 712, on the feedwater line to the reactor, at level +27 (89 feet). They were being helped by three people from phase 1: Nekhayev, the senior mechanical engineer; Uskov, the senior reactor operations engineer; and Orlov, the deputy head of the reactor section. Babichev said I ought to go over there and replace them as they were in bad shape.

Lyutov, the deputy chief scientific engineer, was sitting there with his head in his hands. He kept on repeating that if only someone would tell him the temperature of the graphite in the reactor, he would be able to explain everything. I asked him, in amazement, what graphite he was talking about, and pointed out that practically all the graphite was on the ground. I invited him to take a look, as it was now light outside, and I had just seen it all myself.

Looking very scared and confused, Lyutov asked me what I meant and said he found it hard to imagine.

"Let's go and take a look," I suggested. Together we went along the de-aerator corridor and into the standby control room. Here again, broken windowpanes lay all over the floor and crunched squeakily underfoot. Although I did not realize it at the time, the dense, acrid air, saturated with long-lived radionuclides, was being bombarded with up to 15,000 roentgens per hour of gamma rays from the pile of rubble formed by the explosion. My eyelids and throat were burning, and I was increasingly out of breath. Enormous inner heat radiated from my face, and my skin was dry and tight.

I told Lyutov to look at the graphite strewn all around.

"Are you quite sure it's graphite?" he replied incredulously.

"What else could it be?" I exclaimed indignantly, though I myself was inwardly reluctant to believe what I had seen. Yet I had realized that people were dying needlessly because of all the lies that had been told, and that the time had come to own up.

Further agitated by my own exposure to radiation, I hammered away relentlessly at the same point. "Look! Graphite blocks—you can make out all the details. See, there's one with a male end and another with a female depression; and the holes in the middle are where the fuel channel used to go. Can't you see?"

"Yes, I see. But is it really graphite?" Lyutov continued to doubt my word.

Such blindness always used to infuriate me. People tend to see only what is convenient for them to see—even if it costs them their lives! I began to scream at him, though he was my supervisor, asking him what he thought it could be, if not graphite. Lyutov wanted to know how much of it there was. It seemed, at last, that he was coming to his senses.

I told him that we could only see some of it from where we were standing, and that the explosion must have scattered it in all directions. At seven that morning, from my balcony, I had seen fire and smoke coming from the floor of the central hall.

We returned to the control room. It also smelled of radioactivity in there, and I found myself looking at my familiar No. 4 control room as if for the first time—with its panels, instruments, dials, displays. Everything was dead. The needles of the instrument dials had remained frozen at zero or off the scale. The printer attached to the Skala computer, which in normal operating conditions produced a constant printout of the parameters, was now silent. All of those diagrams and printouts would eventually be important, as the graphs of the technological processes and the figures were all mute witnesses to this nuclear tragedy. It occurred to me that they would soon be cut out and taken like priceless treasures to Moscow, to aid in determining the precise cause of the accident. The logs kept by the operators in the control room and other work stations—now merely a pile of papers in a bag—would also be taken away for scrutiny. Only the 211 round Selsyn dials, showing the positions of the absorber rods, stood out clearly against the dead background of the control panel, which was lit internally by its alarm lights. The needles of the dials were stuck at 8.2 feet (2.5 m), having failed to reach 14.75 feet (4.5 m).

I left the control room and hurried up the stair-elevator well to level +27 (89 feet), so as to replace Akimov and Toptunov in compartment 712. On the way I bumped into Tolya Sitnikov, who looked really ill, his skin a dark brown from the nuclear tan. He could not stop vomiting. Making a big effort to overcome his nausea and weakness, he said, "I checked everything, as Fomin and Bryuk-hanov told me to. They are convinced the reactor is intact. I was in the central hall, on the roof of V block. There's a lot of graphite and fuel up there. I looked down into the reactor. I think it's been destroyed. It's blazing away. It's hard to believe, but it's true."

The fact that he said "I think" says something of the torment Sitnikov was going through. Even he, a physicist, was unwilling to believe the whole truth. What he had seen was so horrendous that he simply could not believe it himself.

Throughout the history of nuclear power, the "it" he referred to had been feared more than anything else. The fear had been hidden—but "it" had happened anyway.

Sitnikov then continued to stagger down the stairs, and I hurried on up to compartment 712. It had a high threshold, about 14 inches (350 mm); and the entire compartment was flooded with fuel-saturated water over the top of the threshold. Akimov and Toptunov emerged, looking extremely depressed, with swollen dark brown faces and hands. In the medical center, they found that their entire bodies were that same color; clothing doesn't stop radiation. Their lips and tongues were so swollen that they could hardly talk. Besides their pain, which must have been terrible, they were also feeling bewildered and guilty.

"I shall never understand it," said Akimov, barely moving his swollen tongue. "We did everything right. Why did it happen? It really hurts, Vitya. We're really screwed up. We opened all the gate valves we could find. Take a look at the third one on each branch line."

They then went down the stairs, and I entered compartment 712, which was quite small, only 86 square feet (8 m²). A thick pipeline passed through the compartment and divided into two branches, each 8 inches (200 mm) in diameter; on each of these there were three gate valves. It was these that Akimov and Toptunov had opened. Akimov was convinced that the water from the functioning feedwater pump was traveling along these pipes and into the reactor; whereas in actual fact it was going nowhere near the reactor, but merely pouring into the lower compartments, flooding the cable housings and switching gear underground, and making an already bad situation worse.

It is a strange thing, but during those truly weird hours, the overwhelming majority of the operational staff, including me, believed what they wanted to believe, and not what was really happening.

The nonsensical but extremely comforting idea that the reactor was intact mesmerized a great many people here, in Pripyat, in Kiev, and also in Moscow, which sent forth a stream of increasingly rigid and vehement commands: "Feed water into the reactor!"

These commands had a tranquilizing effect, instilling confidence, vitality, and strength in men who were under such terrible stress that from a strictly biological point of view they should have been incapacitated.

The pipeline in compartment 712 was half under water, and that water was giving off about 1,000 roentgens per hour. As there was no electric current in any of the gate valves, the handles had to be turned manually, which took hours. Akimov and Toptunov had spent several hours doing just that and collecting their fatal doses. I checked to see how far they had been opened, and found that two were open on each branch, left and right. I started opening the third, but found that they were both damaged. I started to open them a bit farther. I was in the compartment for 20 minutes, during which time I got a dose of 280 rads.*

Back down in the control room, I replaced Babichev. Besides me, the following people were now present at the control panel: Gashimov and Breus, senior unit control engineers; Sasha Cheranev, the senior turbine control engineer, and his stand-in Bakayev; and the shift foreman of the reactor section, Seryozha Kamyshny. He was now rushing all over the unit, mainly along the de-aerator, trying to disconnect the two left de-aerator tanks, from which water was reaching the destroyed feedwater pump. However, he failed to do so. Those particular gate valves were 24 inches (600 mm) in diameter, and as a result of the explosion, the de-aerator had shifted some 18 inches from the main walls of the reactor building, thus severely damaging the gangways. It was now impossible to control the gate valves, even manually. They tried to get them working again, but people were too sick to go on with such high levels of gamma radiation. Kamyshny had been helped by the senior turbine machinist, Kovalyov, and by Koslenko, a metal worker.

By 9 A.M. the functioning emergency feedwater pump stopped working, and it was just as well that it did, because now the underground areas were no longer being flooded. There was no more water in the de-aerator tanks.

I was on the phone all the time to Fomin and Bryukhanov, and they were constantly in touch with Moscow. The version of events that Moscow was getting from them was that water was being fed into the reactor. Back came the order to keep on pumping water into the reactor. But they had run out of water.

Radiation levels in the control room were running at 5 roentgens per hour, but higher in areas exposed to the vast pile of rubble. But there were no instruments, so no one really knew. When I told Fomin that there was no water, he panicked and started yelling, "Get water in there!" Where was I going to get water from?

Fomin was frantically trying to figure out a solution, and eventu-

---

*Rad means "radiation absorbed dose" and indicates the amount of radiation absorbed by a tissue or an organ of the body.—Ed.

ally he came up with one. He sent the chief engineer for the new units, Leonid Konstantinovich Vodolazko, and the unit shift foreman, Babichev—the man I had taken over from—to arrange for water to be fed into the three clean condensate tanks, each of which had a capacity of 265,000 gallons (1,000 m³), the idea being that from there it would again reach the reactor. Luckily Fomin's little stunt did not work.

Around 2 P.M. I left the control room of No. 4 unit. I was already feeling terrible, with vomiting, headaches, dizziness, fainting spells. I washed and changed in the lock-chamber in phase 2 and then went to the laboratory wing of phase 1, to the first-aid post, where there were already doctors and nurses.

Much later on 26 April, new fire crews which had just arrived in Pripyat pumped water containing nuclear fuel from the plant's underground cable housings into the cooling ponds: As a result, the radioactivity of nearly 6,000 acres (22 km²) of water rose to one microcurie (10^{-6} curie) per liter, which is the level found in the main circuit of an operating nuclear reactor.

As we have seen, Fomin and Bryukhanov refused to believe Sitnikov when he said the reactor had been destroyed. They also refused to believe the plant's senior radiation protection official, Vorobyov, who warned them of high radiation levels. Instead, they told him to toss his radiometer in the garbage. However, somewhere deep inside Bryukhanov, a single sober thought had implanted itself. At some subconscious level, he had taken note of the information supplied by Sitnikov and Vorobyov and, as a precaution, requested permission from Moscow to evacuate the town of Pripyat. Even from 2,000 miles away in Barnaul, where his adviser L. P. Drach had reached him by phone, Shcherbina issued clear orders: "Don't start a panic! There must be no evacuation until the government commission gets there!"

Nuclear euphoria, as well as the tragic, catastrophic nature of the situation, had made Fomin and Bryukhanov mentally unstable. Every hour Bryukhanov had been reporting to Moscow and Kiev that the radiation situation in Pripyat and around the nuclear power station was within normal limits, that the situation was generally under control and cooling water was being fed into the reactor.

When the feedwater pump stopped, Fomin frantically tried to find other sources of water.

As V. G. Smagin has testified, Fomin sent Vodolazhko (deputy chief engineer for No. 5 reactor, then under construction) and Babichev, the unit shift foreman, who had not yet managed to get to the medical center, to arrange for firefighting water to be delivered to the three clean condensate tanks, each with a capacity of 265,000 gallons (1,000 m³), which were situated outside, next to the reactor section's auxiliary systems, a short distance from the radioactive pile of rubble. The idea was that from there the emergency pumps would transfer the water from the ECCS system to what Fomin imagined was still the reactor. This fanatical stubbornness, strikingly similar to the crazed single-mindedness of a maniac, could only do more harm, by increasing the flooding of compartments below ground and exposing more and more people to high doses of radiation. The whole of No. 4 reactor building had now been cut off from the grid, the switching gear was flooded with water, and it was no longer possible to turn on any piece of machinery, as it would have exposed the staff to extremely high doses of radiation. The ambient radiation fields ranged from 800 to 15,000 roentgens per hour, although the instruments on hand could not register radioactivity above 4 roentgens per hour.

More than one hundred people had already been sent to the medical center. Now was the time for some sober judgment—but no, Fomin and Bryukhanov persisted in their madness, proclaiming that the reactor was intact and ordering the staff to feed water to it.

FIRST ALARMS

Early on the morning of 26 April, the first group of experts prepared to leave Moscow's Bykovo airport on a special flight for Kiev, and then on to Pripyat. Boris Yakovlevich Prushinsky, chief engineer of Soyuzatomenergo, spent the night on the phone rounding up the members of the group.

A second, higher-ranking group was also preparing to leave Moscow. It consisted of representatives of the Central Committee and the government; the senior assistant to the public prosecutor of the USSR, Yu. N. Shadrin; the chief of Soviet civil defense, Lieutenant General V. P. Ivanov; the commanding officer of Soviet chemical warfare troops, Lieutenant General V. K. Pikalov; as well as ministers, academicians, and high-ranking military officers. This group's special flight was to have left for Kiev at 11 A.M. on 26 April 1986, but difficulties involved in contacting people at the weekend—it was Saturday—delayed their departure until 4 P.M.

Meanwhile, the town of Pripyat, where the staff of the nuclear power station lived, had awakened. Almost all the children went to school.

Testimony of Lyudmila Aleksandrovna Kharitonova, senior engineer in the construction department of the Chernobyl nuclear power station:

On Saturday 26 April 1986, everyone was already preparing for the 1 May holiday. It was a fine warm spring day, and the gardens were in bloom. After work my husband, the head of the ventilation section, had intended to take the children to our cottage in the country, our dacha. All morning I had been doing the laundry and hanging it out to dry on the balcony. By evening it had already collected vast amounts of radioactive fallout.

Hardly anyone among the builders and installers knew anything. Then word came about an accident and fire at No. 4 unit. But what exactly had happened, nobody knew.

The children went to school; the littlest ones played in the street and sand lots and rode their bicycles. By the evening of that day, 26 April, all of them had accumulated high levels of radioactivity on their hair and clothes, but at the time we were unaware of this. Fruit punch was being sold just down the street, and many people were buying it. It was a normal weekend day.

The building workers went to work but were soon sent back home, around midday. My husband also went to work. When he came back for lunch, he told me, "There's been an accident. They won't let us in. The whole plant is cordoned off."

We decided to go to the dacha, *but the militia would not let us through, so we came back home. It's funny, but we still regarded the accident as something separate from our private lives. After all, there had been accidents in the past, but they had concerned only the plant itself.*

After lunch, when they started washing the town, no one took any particular notice. It was a common sight on warm summer days; street-washing machines were nothing special in summer, in fact it was the normal peaceful scene. However, I did notice the white foam along the roadside, but thought nothing of it—maybe it was from the high water pressure.

A group of children from our neighborhood bicycled over to the bridge near the Yanov station to get a good view of the damaged reactor unit. We later discovered that this was the most highly radioactive spot in town, as the radioactive cloud released during the explosion had passed right overhead. But none of this was known until later, and that morning, 26 April, the kids simply wanted to get a look at the burning reactor. They later came down with severe radiation sickness.

After lunch our children came home from school, where they had been warned not to go out into the street and to do their washing at home. Then it began to dawn on us that it was serious.

Different people found out about the accident at different times, but by the evening of 26 April almost everyone knew. Even so, no one got too upset, as all the shops, schools, and offices were working. So we assumed that it wasn't too dangerous.

We began to get more alarmed in the evening. It's hard to say where the alarm came from, perhaps from inside ourselves, perhaps from the air, which by then was beginning to take on a metallic smell. It's hard to say what kind precisely, but it was unmistakably a metallic smell.

In the evening the smell of burning was more pronounced. People were saying that the graphite was burning. The fire could be seen from a long way off, but nobody took any special notice of it.

"There's something burning."

"The firefighters put it out."

"It's still burning anyway."

On the grounds of the plant, 330 yards (300 m) from the destroyed reactor unit, in the office of Gidroelektromontazh, Daniel Terentyevich Miruzhenko, a security guard, waited until 8 A.M. and then, as the head of the division did not answer, decided to walk about a mile to the head construction office and notify Vasily Trofimovich Kizima, the construction manager, or the dispatcher about what he had seen the night before. No one came to take over from him, and no one called to give him instructions. So he locked his office and set off, on foot. He already felt sick and was starting to vomit. In the mirror he saw that he had acquired a deep tan overnight, with no exposure to the sun. On top of that, part of his route to the construction office took him along the path of the nuclear cloud released during the accident.

When he reached the office, he found it closed. There was no one around. It was Saturday, after all.

A stranger standing near the front door said to him, "You'd better get yourself to the medical center and fast. You look terrible."

Somehow, Miruzhenko managed to hobble over to the medical center.

Early on the morning of 26 April, Anatoly Viktorovich Trapikovsky, the driver of the director of Gidroelektromontazh and an avid fisherman, drove to the feeder canal in the official car to catch fry and move on later to pike-perch. Finding his usual route barred by the militia, he turned around and tried to approach the warmwater canal from another direction. But here again, he found the militia in the way. He then cut through the woods, following a barely perceptible path, and reached the canal. The fishermen who had been sitting there since the night before talked about the explosions. They seemed to think that the main relief valves had been triggered, because the noise they had heard sounded like steam being released. But after that, there had been a thunderous explosion, with flames and sparks leaping into the air, and a fireball surging skyward.

The fishermen gradually disappeared. Trapikovsky stayed a while longer, but he was now beginning to be afraid, and he too collected his things and went home.

That same morning, two insulation workers from the night shift at the No. 5 reactor unit then being constructed—Aleksey Dzyubak and his supervisor, Zapyokly—went, after coming off duty, to the chemical protection office over 300 yards from No. 4 unit. On the way they walked through the path of the radioactive cloud, thus coming into contact with the nuclear particles that it had scattered on the ground below, and that were now emitting up to 10,000 roentgens per hour. The total dose received by each man was around 300 rads. They spent six months in No. 6 clinic in Moscow.

Klavdia Ivanovna Luzganova, aged fifty, was on duty at No. 2 entrance on the night of 25–26 April, at the construction site for the spent nuclear fuel depository, over 200 yards from the damaged reactor unit. She received a dose of 600 rads, and died at No. 6 clinic toward the end of July 1986.

On the morning of 26 April, a crew of building workers arrived at No. 5 unit. They were joined there by Vasily Trofimovich Kizima, a fearless, tough man who was in charge of the construction project. Before going to the site, he had driven around and inspected the pile of rubble surrounding No. 4 unit. Having no dosimeter with him, he therefore had no idea of the dose he had received. He later talked to me about it:

I could tell something was wrong, of course, as my chest was very dry and my eyes were burning. It occurred to me that Bryukhanov must have released some radiation. I looked at the pile of rubble and went on to No. 5 unit. The workers came to ask me how long they could work, and how bad the radiation was. They wanted extra pay for the hazardous conditions. All of us, including me, were coughing heavily. Our bodies were protesting the presence of cesium, strontium, and plutonium. Not to mention iodine-131, which was getting into our thyroids. It was stifling. Nobody had any respirators or any potassium iodide tablets. I called Bryukhanov to find out what was happening. He replied that he was studying the situation. Closer to lunchtime I called again and was told that Bryukhanov was studying the situation. I am a builder, not a nuclear plant operator, but even I could tell that Comrade Bryukhanov did not have the situation under control. Once a marshmallow, always a marshmallow, that's him all over.

At noon I released the workers and sent them home. There was no point in waiting for further instructions from the boss.

Testimony of VLADIMIR PAVLOVICH VOLOSHKO, chairman of the Communist Party Executive Committee of the town of Pripyat:

All day long on 26 April, Bryukhanov kept everybody in the dark by repeating that the radiation situation in the town of Pripyat was normal. He really seemed to be cracking up and went about with a distinctly clueless, lost look on his face. As for Fomin—well, when he wasn't giving orders, he was crying and whining, with not a trace of his former arrogance, spitefulness, and confidence. By evening they had more or less pulled themselves together, just as Shcherbina was about to arrive, as if he could save them. Bryukhanov was very relaxed about the explosion, and that's hardly surprising. All he knew about was turbines—in fact he had surrounded himself with turbine engineers, while Fomin hired electrical engineers. Can you imagine? Bryukhanov would send a report on the radiation situation once every hour, saying that everything was normal, that there had been no increase in background levels, and so on. They sent Tolya Sitnikov, a top physicist, right into 1,500 roentgens. It makes me hopping mad just to think of it. Yet when he reported back that the reactor had been destroyed, they wouldn't listen to him.

Of the 5,500 operational staff at Chernobyl, 4,000 disappeared on the first day without saying where they were going.

At 9 A.M. on 26 April 1986, Lidia Vsevolodovna Yeremeyeva, the duty officer at Soyuzatomenergo in Moscow, phoned the construction office at the Chernobyl nuclear power station. In Pripyat, the chief site engineer, V. Zemskov, picked up the phone. Yeremeyeva wanted to know the figures for construction work done in the previous twenty-four hours: the volume of concrete poured, the structural metal parts installed, machinery in use, number of workers at the site of No. 5 unit.

"I think you'd better leave us alone today. We've had a slight accident," replied Zemskov who, having just been around the damaged unit as part of his duties, had been heavily exposed to radiation. Later he started vomiting and was taken to the medical center.

THE EXPERTS CONSULT

At 9 A.M. on 26 April, a YaK-40 took off from Bykovo airport, Moscow, on a special flight. On board was the first interdepartmental expert task force, consisting of B. Ya. Prushinsky, chief engineer of Soyuzatomenergo; Ye. I. Ignatenko, the deputy director of the same organization; V. S. Konviz, the deputy head of Gidroproyekt, an industrial research institute responsible for the overall design of the plant; K. K. Polushkin and Yu. N. Cherkashev, representing NIKIET, the Scientific Research and Design Institute for Energy and Fuel, the main designers of the RBMK reactor; Ye. P. Ryazantsev, representing the I. V. Kurchatov Nuclear Power Institute; and others.

As we have seen, the group had been assembled for this flight by Prushinsky, who phoned each member individually.

On departure from Moscow, the task force had the following scant information, supplied by Bryukhanov, on the matter at hand:

- "The reactor is intact, and is being cooled with water" was very flattering to Polushkin and Cherkashev, as chief designers of the reactor. This news must also have pleased Konviz, as general project manager, since he was the one who had chosen this reliable piece of hardware for inclusion in this nuclear power project.
- "The radiation situation is within normal limits" was reassuring to everyone, particularly to the representative of the I. V. Kurchatov Nuclear Power Institute, Ryazantsev, because the core, which had been based on the calculations done by the institute, had proved to be sound, reliable, and manageable, and the reactor had survived this critical situation.
- "Only two fatal accidents." Not so bad for an explosion.
- "An explosion of gases had destroyed the 29,000-gallon (110-m^3) emergency cooling tank for the motors driving the SUZ (protection and control) system." Well, in the future, some improvement would clearly be needed in the tank's safety features.

At 10:45 A.M. on 26 April, the emergency task force of experts was already in Kiev; and two hours later, cars carrying its members drove up to the Communist Party town committee offices in Pripyat.

Finding out the exact state of affairs was the most urgent piece of business before the group, as it needed to provide the members of the government commission with accurate information as soon as they arrived.

First and foremost, they had to move on to the damaged reactor unit and see everything for themselves, ideally from the air. As it happened, a civil defense helicopter was parked a short distance from the overpass near Yanov station. Some time was spent hunting for binoculars and a photographer with a camera. Binoculars were nowhere to be found, but they did find a photographer. Before taking off, they were still convinced that the reactor was intact and being cooled with water. Between an hour and an hour and a half after the arrival of the task force, the MI-6 helicopter took off, carrying the photographer; Prushinsky, the chief engineer of Soyuzatomenergo; and Polushkin, the chief designer of the reactor. Only the pilot had a dosimeter, which made it possible later to determine the dose of radiation they had absorbed.

They approached from the direction of the cement-mixing complex and the town of Pripyat, in front and slightly to the left of the reactor auxiliary system unit, at an altitude of 1,300 feet (400 m). They went down to 650 feet (200 m) for a closer look. What they saw was profoundly depressing. There was a huge pile of rubble and no central hall. The reactor building was unrecognizable. But they proceeded with a methodical inspection.

"Hover here for a bit," Prushinsky asked.

On the roof of the reactor auxiliary system unit, right next to the wall of V block, they could see heaps of twisted girders, light-colored fragments of walls and ceilings, stainless steel pipework glistening in the sun, black pieces of graphite, and mangled fuel assemblies, brown from corrosion. There was a particularly large amount of fuel and graphite near the square ventilation stack protruding from the roof of the auxiliary unit directly adjacent to the wall of V block. Beyond that, they saw a huge pile of misshapen pipelines, small reinforced concrete structures, hardware, fuel, and graphite, spread out over a radius of about 100 yards, from what had once been the wall of the main circulation pump room along row T, inside the devastated main circulation pump room, whose end wall, toward the liquid and solid waste storage building, now visible on the right, had miraculously remained intact.

This was the pile of rubble under which Valera Khodemchuk lay buried. It was here, too, that Valery Ivanovich Perevozchenko, the reactor shift foreman, absorbed a lethal dose of radiation while looking for Khodemchuk, who had been a member of his shift. It was here, as he clambered over heaps of fallen masonry and hardware, that he shouted plaintively, his voice dry and tense from radiation, "Valera! Answer me! I'm here! Answer me!"

Prushinsky and Polushkin were, of course, unaware of these events. Severely shaken by what they saw, and gradually realizing that something far more terrible than mere destruction had happened, they took careful note of every detail of the scene of devastation below.

All around, on the asphalt, now blue in the sunlight, and on the roof of the liquid and solid waste storage building, they saw pitch-black pieces of graphite and even whole assemblies of graphite blocks. There was graphite everywhere, it seemed.

Prushinsky and Polushkin stared, dumbfounded. Until then they had known in theory that such extreme devastation was possible, and may have vaguely imagined what it looked like. But the reality before them now was far more vivid, complex, and stark. The famous saying of the Russian folk figure Kuzma Prutkov—"Never believe what you see with your own eyes!"—now seemed to be coming true. Both Prushinsky and Polushkin found themselves shrinking away from the scene below, as if it was of concern not to them but only to others, to strangers. But it did concern them. And how undignified it was to be forced to see such things.

At first, it was as if they were not actually seeing the catastrophic mess below, but were merely experiencing a sick thrust of apprehension. The sight was so repugnant that they recoiled instinctively from it. If only there was some way they could avoid looking down! But they had to, it was their duty.

The main circulation pump room seemed to have been destroyed by an explosion from within. But how many explosions had there been? In the pile of rubble which sloped up to the floor of what had once been the separator room, they could see sections of long thick piping, probably headers. One of these stretched out almost at ground level across the corner formed by the walls of the auxiliary unit and the pump room; while the other one was much higher up, roughly from level +12 (40 feet) to level +24 (80 feet), its upper end leaning against a long downcomer. This pipe must have been ejected from the suppression pool well. On the floor—if such a shapeless heap can be called a floor—at level +32 (105 feet), some uprooted 130-ton drum-separators were shining brightly in the sunshine, together with eight severely mangled connector pipes, heaps of assorted debris, and fragments of concrete ceiling and wall panels, now dangling in midair. However, the walls of the separator compartment had been blown away, with the exception of one surviving stump near the central hall. Between that stump of a wall and the pile of rubble was a black, gaping rectangular hole, from which nothing at all could be seen protruding. This meant that it led to the suppression pool well or to the upper communication lines of the reactor. Part of the piping and hardware seemed to have been blasted out of this hole. The absence of protruding structures suggested that there must also have been an explosion in there.

With these thoughts passing through his mind, Prushinsky found himself thinking back to the brand-new circuit the way it had looked just after it had been installed—a superb piece of technology. But now . . .

Where the central hall adjoined the de-aerator, a lower segment of wall was still standing. The end wall of the reactor hall along row T had survived, approximately as far as level +51 (167 feet), to the base of the protection and control system (SUZ) emergency tank, which was mounted in that wall, from levels +51 (167 feet) to 70 (230 feet). According to Bryukhanov, that tank had been the site of the explosion of detonating gas which had destroyed the central hall. But what about the rooms housing the main circulation pumps, the drum-separators, and the suppression pool? What had destroyed them? No! Bryukhanov's version was mistaken, perhaps even deliberately misleading.

One particularly telling feature of the scene was the presence of the black pieces of the graphite stack, which lay strewn around the pile of rubble. The fact that there was graphite on the ground led inescapably to one conclusion.

Even now, it was difficult to face the simple and obvious truth: the reactor had been destroyed.

Such acknowledgment would instantly raise the issue of some-one's colossal responsibility to others—indeed, to millions of people, to the entire world. As well as the fact of an unspeakable human tragedy.

That made it easier to look rather than to think. First, one had to try to grasp the full magnitude of the nightmare of a dying reactor, overflowing with poisonous radiation.

The wall of V block toward the auxiliary unit, where it adjoined the main circulation pump room, was torn and jagged. Square pieces of graphite with holes in the middle, formerly part of the stack, could be seen quite clearly on the roof of V block. With the sun high in the cloudless sky, and the helicopter poised 490 feet (150 m) above the roof of V block, there could be no mistaking it. There were piles of graphite closer to the end wall of V block. Pieces of graphite were also evenly strewn around on the roof of the central hall of No. 3 unit and also on the roof of V block, on which the ventilation stack, painted white with red horizontal stripes, was located. Graphite and fuel could also be seen on the inspection platforms of the ventilation stack. And all of these objects were sources of intense radiation. There, below them, was the roof of the de-aerator, where, only seven hours before, the firefighters under Major Telyatnikov had finished battling with the flames.

The flat roof of the turbine hall—now bristling with twisted metal reinforcing bars and shattered metal trusses, all black from the flames—had been blown inside out. The streams of molten tar, in which the firefighters had toiled up to their knees during the night, had now solidified. Tangled fire hoses and hose-reels lay strewn about on those sections of the roof that were still intact.

Abandoned fire trucks, like little red boxes, could be seen along the end wall of the turbine hall, on the corners of rows A and B, and also along the intake basin—silent witnesses to a tragic struggle between frail humans and the massive power, both visible and invisi-ble, unleashed by the atom.

Over to the right, looking like children's sandals from that altitude, a number of sailboats and motor boats lay on the golden sand beaches; and beyond them the vast smooth expanse of the cooling pond reservoir, where the water was still clean.

Some stragglers could still be seen leaving the construction site of No. 5 reactor, individually or in groups. These were workers who had been released some time previously by Kizima, the site supervisor, who had decided not to wait for Bryukhanov to tell him the truth. All of them were to be exposed to radiation as they walked through the path of the radioactive cloud, and all of them were to take this terrible poison home with them on the soles of their shoes.

"Hover right above the reactor," said Prushinsky to the pilot. "That's right! Hold it there! Take some pictures!"

The photographer did his bidding. They opened the door and stared down. The helicopter was now in the middle of the radioactive plume rising from the reactor, yet there were no respirators and no radiometers on board. The black triangle of the spent fuel storage pool was now below them, but they could see no water in it.

"The fuel in the basin will melt down," thought Prushinsky.

And there was the reactor—the round "eye" of the reactor well. It seemed to be half shut. The huge "eyelid," the upper biological shield, was out of position and cherry-red from the intense heat. Flames and smoke were rising from the half-closed "eye." It looked just like a gigantic ear of barley which was about to ripen and burst.

"Ten ber," said the pilot, glancing at the dial of the optic dosimeter. "Today we'll have to do this several times."

"Get us out of here!" Prushinsky commanded.

The helicopter descended from its vantage point above the central hall and headed for Pripyat.

"Well, it's all over," glumly remarked Konstantin Prushinsky, representative of the main designer of the reactor.

After their aerial survey of the damaged reactor unit, they drove to see Bryukhanov in the bunker. Both he and Fomin looked extremely downcast. His first words to Prushinsky had a tragic ring: "That's it. The reactor unit is finished," he said, in a subdued voice.

Prushinsky, however, could still hear Bryukhanov summarizing the accident the night before: "The SUZ emergency water tank has exploded. The roof of the central hall is partly destroyed. The reactor is intact. We are feeding water into it."

"Well, then, which is it? Is the reactor intact or not?" Prushinsky said to himself.

He and Bryukhanov got into the car and drove around the destroyed unit one more time.

Testimony of LYUBOV NIKOLAYEVNA AKIMOVA, the wife of the shift foreman of No. 4 unit:

My husband was a very nice and very sociable person. He got on well with people, but without being familiar. He was generally a very happy, obliging person. He was very involved in civic affairs and was a member of the Communist Party town committee of Pripyat. He was tremendously fond of his sons and very concerned. He liked to go hunting, especially when he started working at the reactor unit and we bought a car.

We arrived in Pripyat in 1976 after finishing the Moscow energy institute. First, we worked with the project planning group of Gidroproyekt. In 1979 my husband joined the operational staff. He was senior turbine control engineer, senior unit control engineer, shift foreman in the turbine section, and deputy shift foreman for the unit. In January 1986 he became shift foreman for the unit. This was the post he held when disaster struck.

On the morning of 26 April, he failed to come home from work. I phoned him at the No. 4 unit control room, but there was no answer. I also called Bryukhanov, Fomin, and Dyatlov, but there was no answer. Much later I discovered that the phones had been disconnected. I was very upset. All morning long I ran around, asking everybody and trying to find my husband. They already knew that there had been an accident, and I then got even more worried. I rushed to see Voloshko, at the Pripyat municipal executive committee, and Gamanyuk, at the party town committee. Eventually, after asking a lot of people, I found out that my husband was in the medical center, so I went there immediately. But they told me that he was having an intravenous infusion, and wouldn't let me see him. Instead of going away, I went up to the window of his room. He soon came to the window. His face was dark brown.

When he saw me, he laughed. He was overexcited; he tried to calm me down, asking me about our sons through the closed window. I got the impression that just then he felt particularly glad that he had sons. He told me to keep them indoors. He was even cheerful and I was somewhat reassured.

Testimony of L. A. KHARITONOVA:

Toward evening on 26 April, someone started a rumor, to the effect that those who wanted to could evacuate themselves in their own cars. Many people left in their cars that same day for different parts of the country.*

But we were evacuated in the evening on 26 April on the Khmelnitsky-Moscow train. Soldiers were patroling at the Yanov station. There were lots of women with small children. They all looked a bit confused, but they behaved calmly. People were looking at the soldiers inquisitively, as if they were searching for any sign of alarm or fear. But the soldiers were quite relaxed and friendly, all smiling. The trouble is, the radioactive cloud had passed right over Yanov; there was heavy radiation on the ground, on the trees, everywhere. But nobody knew about it at the time; everything looked perfectly normal. But I nonetheless felt a new age had dawned. And when the train pulled in it seemed to me so very different, as if it had just come from the old, clean world we used to know, into our new poisoned age, the age of Chernobyl.

The stewardess was heating water in the coach. We washed the little girl. We put her clothes in a plastic bag and shut it up in the suitcase. And off we went. All the way to Moscow we washed and washed our things. And we took our worries and our pain farther and farther away from Pripyat.

Testimony of G. N. PETROV, former head of the equipment section at the Pripyat branch of Yuzhatomenergomontazh:

We woke up at 10 A.M. on 26 April. It was just a day like any other. The warm light of the sun was reflected on the floor, and a blue sky outside. I felt really good. I was back home and intended

*Taking the radioactive poison with them on their clothes and on the tires of their cars.

to take it easy. I went out onto the balcony for a smoke. There were lots of children in the street. There were kids playing in the sand, building houses, making pies. The older ones were racing about on their bikes. Young mothers were out pushing their baby carriages. Everything looked normal. And all of a sudden I remembered driving up to the unit the night before. I had been really scared. Now I also remember being puzzled. How could such a thing happen? Everything was normal, but at the same time, everything was terribly radioactive. It took me some time to feel bad about the poisons I couldn't see, because everything was just as usual. You look, and see everything is clean, but really it's filthy. It's hard to grasp such things.

By lunchtime I was feeling cheerful. There was now a sort of sharpness about the air. Not exactly a metallic taste in the air, but something tart, and a funny acid taste around my teeth, as if I had just licked a weak battery to check it out.

Around eleven o'clock, our next-door neighbor, Mikhail Vasilyevich Metelev, an electrical assembly man with Gidroelektromontazh, went up onto the roof and stretched out on a rubber mattress to work on his tan. At one point he came down for a drink and said how easy it was to get a tan that day, he had never seen anything like it. He said his skin gave off a smell of burning right away. And he was tremendously jolly, as if he just been boozing. He invited me to join him, but I did not go. He said, "Who needs to go to the beach?" And you could see the reactor burning, in the background, quite clearly, against the blue sky.

As I later discovered, there was as much as 1,000 millibers per hour in the air at that time. Plutonium, cesium, strontium, and, of course, plenty of iodine-131: By evening it had really lodged in our thyroids. All of us, children and adults.

But we knew nothing then. We were just living our normal lives—happy lives, as I remember.

That evening our neighbor who had been tanning himself on the roof began vomiting uncontrollably and was taken to the medical center. After that he went to Moscow, or maybe Kiev. But I thought nothing of it at the time, because it was a normal summer day, with the sun shining, a blue sky, warm weather. Even on days like that, *people do get sick every now and then, and get taken away in the ambulance.*

Everything about that day was normal. It was only later, when we were told everything, that I thought back to the night when I drove up to the plant. I remember the potholes lit up by my head-lights, and the cement factory all covered in dust. For some reason it stuck in my mind. And now it occurs to me that pothole was radioactive, a perfectly ordinary pothole, and the entire cement factory and everything else—the sky, my blood, my brain, and what I was thinking—all was radioactive. Everything.

Meanwhile, at Bykovo airport in Moscow, the members of the government commission were preparing to take off. Although their special flight had been scheduled for 11 A.M., it took some time for everybody to assemble. Departure was postponed, first to 2 P.M., and then to 4 P.M.

Testimony of MIKHAIL STEPANOVICH TSVIRKO, director of the All-Union Nuclear Power Construction and Assembly Enterprise, Soyuzatomenergostroy:

On the morning of 26 April 1986, I had high blood pressure and a headache, so I went to the polyclinic of the No. 4 department of the USSR Ministry of Health.*

Around 11 A.M. I phoned the office to see how our various con-struction sites were doing. The duty officer was Lidia Vsyevolo-dovna Yeremeyeva, the senior specialist of the technical department. She said that the Chernobyl site had not given her the usual sum-mary. The chief engineer or the dispatcher had told her that they had had an accident, and Kizima had let his people from No. 5 unit go home.

Yeremeyeva also told me that Minister Mayorets had been look-ing for me.

I phoned the minister's deputy. He sounded very nervous, saying that he had been unable to find me at home or at work and told me

---

*The Kremlin hospital.

to pack immediately and go to Bykovo airport, where I would catch a flight to Chernobyl. I packed and went to Bykovo. Aleksandr Nikolayevich Semyonov was already there, pacing up and down. He told me that four roof girders had collapsed in the turbine hall, where it adjoins No. 4 unit.

"Is there any contamination?" I asked.

"No, there's no contamination," he answered. "Everything is clean."

I was already starting to figure out which cranes would have to be moved in to put the girders back in place, when up came V. V. Marin, the senior man in charge of nuclear energy matters in the Central Committee. He told me that, besides the fallen girders in the turbine hall, the roof over the reactor had also collapsed.

"Any contamination?" I inquired.

"Amazingly there is no contamination," said Marin. "And the main thing is that the reactor is intact. It's a really terrific reactor! Dollezhal really did a great job designing such a machine!"

The job was getting more complicated, and now I started to think how I would get my cranes into the central hall.

Here I shall interrupt the testimony of M. S. Tsvirko, with whom I worked side by side for four years at Soyuzatomenergostroy.

Mikhail Stepanovich Tsvirko, an experienced, tenacious businesslike man who spent decades building factories for different ministries, and a former head of Glavzavodspetsstroy (the central directorate for special industrial construction), was practically compelled to take the top job at Soyuzatomenergostroy, which until then had persistently failed to meet its targets. It was to his credit that he was strongly opposed to this move, arguing that he knew nothing about nuclear power stations, and that nuclear energy was an alien, incomprehensible field. But the people in the nuclear power office of the Central Committee and the minister himself did some arm twisting, and he consented. For him, respect and obedience for his superiors were the rule. In fact Tsvirko not only respected his superiors but was scared stiff of them, and did not hide the fact.

As Tsvirko was unfamiliar with the details of nuclear power plants, he concentrated on planning targets. He knew how to count money. We, his subordinates, did all the thematic studies, and taught our new boss the mysteries of nuclear power as we went along.

He was short and squat, a bit on the heavy side. In his youth he had been a boxer and still had the characteristic flattened boxer's nose. He was bald, kept his square jaws tightly clenched (a boxer's habit), and had the slightly slanting eyes of a Tatar warrior. This was the man who, within one year, had brought the whole enterprise back on target.

Still, he was terrified all the time. People laughed at him behind his back, but I felt that Tsvirko suffered not so much from fear of his superiors as from pangs of conscience. He was always worried that we were not working as well as we should. In those days, whenever some particular assignment was behind schedule, he often said to me, "They'll slay us, I tell you! Nobody's going to listen to anything we have to say."

He also admitted that he feared radiation more than anything else, because he understood nothing about it.

And he was now at Bykovo airport.

The special flight took off for Kiev at 4 P.M. The following passengers were on board: Yu. N. Shadrin, senior assistant to the attorney general of the USSR; A. I. Mayorets, minister of energy and electrification of the USSR; V. V. Marin, the nuclear power secretary of the Communist Party Central Committee; A. N. Semyonov, deputy minister of energy; A. G. Meshkov, the first deputy minister of medium machine building; M. S. Tsvirko, director of Soyuzatomenergostroy (the Department of Nuclear Power Plant Construction); V. A. Shevelkin, deputy director of Soyuzelektromontazh; L. P. Drach, adviser to B. E. Shcherbina; Ye. I. Vorobyov, deputy minister of health of the USSR; and V. D. Turovsky, deputy director of the ministry of health of the USSR.

They sat in the lounge of the YaK-40 on red couches facing each other. Marin (he had been phoned by Bryukhanov at 3 A.M.) repeatedly shared his thoughts with the members of the government commission: "The one thing I was particularly happy about was that the reactor withstood the explosion of the SUZ tank. Academician Dollezhal did a great job! This is a superb nuclear reactor, and he

153

was the man in charge of building it. Bryukhanov woke me up with his phone call at 3 A.M. and told me that there had been a terrible accident, but that the reactor was intact and cooling water was being fed into it."

"You know, Vladimir Vasilyevich," said Mayorets, a man whose only field of expertise was transformers and who was totally out of his depth in the fast-paced world of nuclear power, "I don't think we're going to be sitting around in Pripyat for long."

He said the same thing an hour and a half later, on the AN-2 which was taking the members of the government commission from Zhuliana airport to Pripyat. The Ukrainian minister of energy, V. F. Sklyarov, was traveling with him. When he heard Mayorets's optimistic forecasts about the length of their stay in Pripyat, he corrected his superior: "Anatoly Ivanovich, I don't think two days is going to be enough."

"Don't try to frighten us, Comrade Sklyarov," Mayorets replied brusquely. "Our main job is to restore the damaged reactor unit as soon as possible and get it back on the grid."

Testimony of M. S. Tsvirko:

When we landed in Kiev, the Ukrainian minister of energy, Sklyarov, told us right away that there was radiation in Pripyat. I personally found that alarming, as I felt sure that something must have happened to the reactor. But Minister Mayorets was quite calm.

The worst of it was that for three nights we slept at the Pripyat Hotel, where the contamination was pretty bad.

At approximately the same time as the members of the government commission were landing in Kiev, the personal aircraft of the deputy chairman of the Council of Ministers of the USSR, B. Ye. Shcherbina, was approaching Moscow from Barnaul. On arriving in the capital, Shcherbina changed, had dinner, and left from Vnukovo airport for Kiev; he arrived in Pripyat around 9 P.M.

Testimony of G. A. Shasharin:

Mayorets flew in. We went on board an AN-2 and took off for Pripyat. On the way I told Mayorets of the need to set up working groups at the site of the accident. I had thought this over previously, on the flight from Simferopol to Kiev. In my opinion, such groups should help carry out the work of the government commission in an orderly manner, and help in the preparation and adoption of decisions. Here is the list of groups I suggested to Mayorets:

A group to study the causes of the accident and safety at nuclear power plants—Shasharin, Meshkov.

A group to study the actual radiation situation around the plant—Abagyan, Vorobyov, Turovsky.

A group for emergency restoration work—Semyonov, Tsvirko, installers.

A group to assess the need for the evacuation of the population of Pripyat and nearby farms and villages—Shasharin, Sidorenko, Legasov.

A group to supply instruments, equipment, and materials—Glavenergokomplekt (the department supplying energy-related equipment to the facilities of the Ministry of Energy and Electrification) and Glavsnab (the department of supply).

The plane landed at the small airstrip between Pripyat and Chernobyl, where cars were waiting. We were met by General Berdov; Gamanyuk, the secretary of the local party town committee; Voloshko, the chairman of the local party executive committee; and others. Marin and I got into a small Gazik automobile, driven by Kizima, and asked him to go straight to the damaged reactor unit. Mayorets was also anxious to go there, but was dissuaded from doing so; instead, he went with the rest of the team to the Communist Party town committee offices.

We passed through the police cordon and turned into the plant grounds.

I shall now interrupt the testimony of G. A. Shasharin, in order to describe V. V. Marin, the man in charge of nuclear power matters in the Communist Party Central Committee.

By training and work experience, Vladimir Vasilyevich Marin was an engineer specializing in the construction of nuclear power stations. For many years he worked as the chief engineer of a construction and assembly firm in Voronezh and was involved in work on the Novo-Voronezh nuclear power station. In 1969 he was invited to work in the Communist Party Central Committee as an instructor to the committee on nuclear power in the engineering division.

I saw him quite often at board meetings of the Ministry of Energy of the USSR, at party meetings, and at reviews and appraisals of the work of individuals in the major departments and branches of the nuclear power industry. Marin took an active part in the work of start-up teams at the various nuclear construction sites and was personally acquainted with the construction managers of all nuclear power stations. Moreover, when they ran into difficulties he helped provide them with equipment, supplies, and labor, bypassing the Ministry of Energy.

Physically he was big, with reddish hair, a booming bass voice, and distinctly shortsighted, with thick horn-rimmed spectacles which glinted in the light. I found his frankness and clear-sightedness most appealing. Marin was a hardworking, dynamic, capable engineer who was always striving to improve his professional qualifications. Nonetheless, he was primarily a builder and out of his depth when it came to the operation of nuclear power stations.

Late in the 1970s, while I was department head at Soyuzatomenergo, I often met him at the Central Committee where he was at the time the only person dealing with nuclear power issues.

When discussing business, he usually digressed in a rather lyrical mode, complaining that he was overworked. "You have ten people in your department, whereas I have to handle the whole of the country's nuclear power industry," and he would then say, pleadingly, "I need your help right away. Help me get material, information."

About that time he had frequent spasms of the cerebral blood vessels, with fainting spells, and doctors had to come to his home on emergency visits.

Early in the 1980s, a nuclear power sector was organized in the Central Committee of the party, with Marin at its head; then, at last, he was provided with some assistants. One of them was G. A. Shasharin, an experienced nuclear power man who had for many years been involved in the operations of nuclear power stations, and later became the deputy minister of energy for the operation of such plants.

Marin was now driving with him in Kizima's Gazik on the way to the destroyed unit. As they drove along the Pripyat-Chernobyl highway, they saw buses and private cars going the other way. Spontaneous evacuation had already begun. A number of people with their families and their radioactive possessions had been leaving Pripyat during the daytime on 26 April, without waiting for instructions from the local authorities.

Testimony of G. A. SHASHARIN:

Kizima drove us to the end of No. 4 reactor building. We got out of the car right next to the pile of rubble, with neither respirators nor protective clothing. No one among the new arrivals was aware of the extent of the disaster, and Bryukhanov and Fomin could not have cared less about us. We had difficulty breathing, our eyes smarted, we were coughing severely and, deep inside, we felt extremely worried and vaguely anxious to get out of there and go somewhere else. Another thing—it was really upsetting to look at all that. We found ourselves wondering whether we were the ones who had brought all "that" about. When the destroyed unit first came into view while we were still on the road, Marin started cursing and swearing. "Look at the mess we've landed ourselves in. That's just great. Now we're all lumped together with Bryukhanov and Fomin."

Kizima had been at the plant since morning. Needless to say, we had no dosimetrists with us. All around us there were fragments of fuel and graphite. The drum-separators, which had been knocked off their supports, were shining in the sun. A faint black smoke was rising from a halo of flames rather like a solar corona above the floor of the central hall, from somewhere near the reactor. At the time we thought that something was burning on the floor; it never occurred to us that it was the reactor. Marin was beside himself with rage, cursing, and gave a graphite block a hearty kick. We did not realize that the graphite was emitting 2,000 roentgens per hour—and the fuel, 20,000. The partly crushed emergency tank was plainly visible, so it was obvious that the explosion must have occurred somewhere else. Without the slightest fear, Kizima was walking about, looking very much in charge, and lamenting the fact that after all the effort that had gone into building the place, now they were walking around among the wreckage of the fruits of their labors. He said that he had come out a number of times since morning to make sure the whole thing wasn't some kind of mirage. It had turned out that it was not a mirage. He even pinched himself a couple of times. He said that he had never expected anything else from that marshmallow Bryukhanov. In the opinion of Kizima, it was bound to happen sooner or later.

We drove around the plant and then went down into the bunker, where we found Prushinsky, Ryazantsev, Fomin, and Bryukhanov. Bryukhanov was extremely sluggish and apathetic, just staring off into the distance; but he responded to commands briskly and precisely. Fomin, on the other hand, was extremely agitated, with a crazy glint in his inflamed eyes, as if he was just about to burst; but he did everything quickly. Then he slumped into a deep depression.

While still in Kiev, I had asked Bryukhanov and Fomin whether the pipelines feeding water to the reactor were intact, and also the downcomers from the drum-separators to the headers. They had assured me that the pipes were all intact. It then occurred to me that a solution of boric acid ought to be fed into the reactor. I ordered Bryukhanov to do whatever was necessary to restore current to the damaged unit and connect the feedwater pump so that it would supply water to the reactor. I figured that at least some of the water would reach the reactor. I asked whether they had any boric acid at the plant. He replied that they did, but not very much. I got in touch with suppliers in Kiev, who found several tons and promised to deliver it to Pripyat by that evening. By then, however, it had become obvious that all the pipelines from the reactor had been severed and the acid would not be needed. But we did not realize this until the evening of 26 April. In the meantime, however, feeling quite convinced that the reactor was intact and was being supplied with water, Marin, Kizima, and I drove to the meeting of the government commission in the offices of the Pripyat town committee.

Testimony of Vladimir Nikolayevich Shishkin, deputy director of Soyuzelektromontazh, in the Ministry of Energy of the USSR, a participant in the meeting held at the offices of the party committee of the town of Pripyat on 26 April 1986:

We all met in the office of A. S. Gamanyuk, first secretary of the Party Committee of the town of Pripyat. The first speaker was G. A. Shasharin, who said that the situation was serious but controllable. Cooling water was being fed into the reactor. They were trying to find some boric acid, and very soon would be pouring a solution of that acid into the reactor, in order to extinguish the fire immediately. Admittedly, there was some reason to believe that not all the water was actually reaching the reactor: the underground cables and the switching gear were flooded, some of the pipes were broken, and it was possible that part of the reactor might have been damaged. In order to find out the exact state of affairs, Fomin, Prushinsky, and physicists from the nuclear power institute had gone again to see what had happened to the reactor. They were expected back at any minute to report on their findings.

Shasharin must have guessed that the reactor had been destroyed; he had seen fragments of graphite and fuel on the ground, but he just did not have it in him to say so. At least, not right away. He would have to work up to it gradually, so that the mind would be able to come to terms with the ghastly, catastrophic reality of what had happened.

Shasharin went on to say that the representative of the general designer had been sent over there to take a look. He felt that the time had come for a collective effort. No. 4 unit now had no electrical current at all. The transformers had been disconnected to guard against short circuits. The underground cable ducts were flooded between units 4 and 1. As the underground switching gear was flooded, he had instructed the electricians to find 2,300 feet (700 m) of power cable and have it on hand. Fomin and Bryukhanov had also been instructed to make sure that electrical, water, and other communication lines to the damaged reactor unit were kept quite separate from those connected to the units still in operation. Lelechenko, the deputy head of the electrical section, was in charge of the electricians.

"What kind of a plant design is that?" said Mayorets indignantly. "Why weren't those lines kept separate in the original design?"

"Anatoly Ivanovich, I'm talking about what's actually happening. As to why that's the way it is, that's another matter. In any case, we're trying to find a cable, water is being fed into the reactor, and the communication lines are being kept separate. It looks as if radiation levels are high all around No. 4 unit."

With his booming bass voice Marin interrupted Shasharin and spoke straight to the minister: "Anatoly Ivanovich! A moment ago Gennady Aleksandrovich and I were near No. 4 unit. It's horrible. It's hard to believe that it's come to this. There's a smell of burning, and graphite all over the place. I even kicked a graphite block to be sure what it was. Where did the graphite come from? How much of it is there?"

"I've been wondering about that, too," said Shasharin. "Maybe some of the graphite was blown out of the reactor. Some of it—"

"Bryukhanov!" said the minister, turning to the plant director. "You've been telling us all day long that the radiation situation was normal. What about the graphite?"

Bryukhanov, his face ashen, his eyelids puffed and red, rose slowly as usual and remained silent for a long time. He always waited a long time before saying anything, but now he had plenty to think about. In a subdued voice he said, "It's hard to imagine. The graphite we received for the new No. 5 unit is all there, intact. I originally thought it must be that graphite, but it's all there. In that case it might have been ejected from the reactor. Some of it, anyway. But then . . ."

"It seems radiation is high all around the unit," said Shasharin, returning to his point. "It's not possible to take accurate readings. We don't have radiometers with the right range. The ones we have go up to 1,000 microroentgens per second, which is 3.6 roentgens per hour. In that range, the instruments are off the scale wherever you go. We assume the background radiation is very high. Actually there was one radiometer, but it was buried under rubble after the blast."

"Outrageous!" Mayorets muttered. "Why don't you have the instruments you need?"

"The original design made no provision for such an accident.

What happened was unthinkable. We have asked for assistance from national civil defense and chemical warfare units. They should be here soon."

"What exactly happened?" asked Mayorets. "What was the cause?"

"We're still not sure," said Shasharin. "It happened at 1:26 A.M. during an experiment with the inertial rundown of the generator."

"The reactor must be shut down right away!" said Mayorets. "Why have you got it running?"

"The reactor has been shut down, Anatoly Ivanovich," said Shasharin.

It seemed that all those responsible for the disaster were anxious to delay as long as possible the awful moment of reckoning, when the truth would be disclosed in all its details. They were really acting as they had always done at Chernobyl, hoping that the bad news would announce itself, and that responsibility and blame would somehow be spread imperceptibly and quietly over everyone. This accounts for their studied slowness, at a time when each minute was precious, when delay caused the criminal exposure of innocent residents to radiation, and when each one of them was nervously poised to utter the word evacuation, but . . .

"That's not all, Anatoly Ivanovich," Shasharin replied. "The reactor is now in an iodine well, which means that it is thoroughly poisoned."

Meanwhile, the reactor was burning away. The graphite was burning, belching into the sky millions of curies of radioactivity. However, the reactor was not all that was finished. An abscess, long hidden within our society, had just burst: the abscess of complacency and self-flattery, of corruption and protectionism, of narrow-mindedness and self-serving privilege. Now, as it rotted, the corpse of a bygone age—the age of lies and spiritual decay—filled the air with the stench of radiation.

Testimony of V. N. SHISHKIN:

One got the general impression that those chiefly to blame— Bryukhanov, Fomin, Meshkov, Kulov, and others—were down- playing what had happened and the hazards involved.

We then heard from A. S. Gamanyuk, first secretary of the Party Committee of the town of Pripyat, who at the time of the accident had been in the medical center for a checkup. On the morning of 26 April, once he heard what had happened, he walked out and went to work.

Addressing Mayorets, he said, "Anatoly Ivanovich, despite the difficult and even serious situation at the damaged reactor unit, everything is calm in the town of Pripyat, with no panic or disorder. People are just doing whatever they normally do on a weekend. Children are playing in the streets, classes are proceeding as usual in the schools, sporting events are being held. There have even been weddings—in fact, sixteen weddings of members of the Young Communist League have taken place today. We are putting an end to rumors and misleading reports. There have been casualties at the damaged reactor unit. Two operators were killed—Valery Khodemchuk and Vladimir Shashenok. Twelve people were taken to the medical center in serious condition. Another forty people, whose injuries were less serious, were hospitalized later. More patients continue to arrive at the medical center. Bryukhanov, the director of the plant, has been reporting to Kiev and to us every hour that the radiation situation is within normal limits, so we are awaiting instructions from the high-level commission."

The next speaker was a calm and composed man, Gennady Vasilyevich Berdov, a tall, gray-haired major general in the Ministry of Internal Affairs (MVD), deputy minister of internal affairs of the Ukrainian SSR. He had gone to Pripyat at 5 A.M. on 26 April in a brand-new uniform, with gold braid, a brightly colored array of ribbons, and a badge testifying to his record in the Ministry of Internal Affairs of the USSR. Yet his uniform and his gray hair were already heavily contaminated, as the general had spent the early morning hours next to the nuclear power station. In fact, the hair and clothes of all those present, including Minister Mayorets, were also radioactive. Like death, radioactivity does not distinguish between ministers and ordinary mortals. It settles on anybody who happens to be in its way and penetrates their bodies—but none of those at the meeting were aware of this. No protective clothing or instruments to measure radiation doses had been issued to anyone. After all, Bryukhanov had been telling everybody that the radiation *situation was normal; that being so, who needed protective clothing and instruments?*

Turning to Mayorets, General Berdov said, "Anatoly Ivanovich, at 5 A.M., I was already in the vicinity of the damaged reactor. Detachments of militia literally took over from the firefighters. They sealed off all the roads to the plant and the workers' settlement. The countryside around the plant is really nice, so people like to go there on their days off; and today, a Saturday, is one such day. But these recreational areas have now become a danger zone, although Comrade Bryukhanov has been telling us that the radiation situation is normal. On my orders, militia detachments closed off access to these areas, particularly the places on the banks of the cooling pond, and the intake and outflow channels, which are popular fishing spots.* An emergency headquarters was set up in the Pripyat militia station, with reinforcements coming from Polesye, Ivankov, and Chernobyl local militia forces. By 7 A.M. more than one thousand members of the MVD had arrived in the area of the accident. Reinforced detachments of transport militia were mobilized at the Yanov railroad station. At the time of the explosion, freight trains carrying extremely valuable equipment were standing in the station, and passenger trains were coming and going on schedule, their crews and passengers totally unaware of what had happened. Since it is summer, the windows of the trains were open. I'm sure you know the railroad runs within 550 yards (500 m) of the damaged reactor unit. I think the radiation is reaching the cars of those trains. Train movements must be halted.† We have not only sergeants and adjutants on duty but also colonels in the militia. I have been checking our posts within the danger zone. No one has left his post, and no one refused to go on duty. We have been busy contacting the transport depots in Kiev; eleven hundred buses have been moved up to Chernobyl, in case the population needs to be

---

*It is worth noting at this point that General Berdov, while intuitively sensing the danger, did not realize what kind of danger he was dealing with, what the "enemy" looked like, and how he should attack it and defend against it. For that reason his militiamen did their job with neither dosimeters or protective measures, and all of them were exposed to severe doses of radiation. However, they had instinctively done the right thing, by severely curtailing access to the presumed danger zone.

†Once again General Berdov is to be commended, as he was the first of the high-ranking officials present who assessed the situation correctly, even though he had no specialized knowledge of nuclear technology.

evacuated, and they are now waiting for instructions from the government commission."

"What's all this talk of evacuation?" said Minister Mayorets angrily. "Are you trying to start a panic? We have to stop the reactor, and the whole thing will be over. Radiation will return to normal. How is the reactor, Comrade Shasharin?"

"The reactor is in an iodine well, Anatoly Ivanovich," Shasharin replied. "According to Bryukhanov and Fomin, the operators shut it down by pressing the level-5 AZ button. So the reactor is thoroughly shut down."

Shasharin was entitled to say what he did, because he did not know the real condition of the reactor; he had not seen it from the air.

"And where are the operators? Can we invite them in here?" asked the minister, pressing his point.

"The operators are in the medical center, Anatoly Ivanovich. In critical condition."

"I proposed evacuation early in the morning," said Bryukhanov in a subdued voice. "I asked Comrade Drach in Moscow, but I was told to do nothing about it until Comrade Shcherbina got here. And I was not to start a panic."

"Who has inspected the reactor?" Mayorets inquired. "What state is it in now?"

"Prushinsky and the representative of the chief designer of the reactor, Polushkin, have inspected the reactor from a helicopter. They took some pictures and should be here any minute."

"What does Civil Defense have to say?" asked Mayorets.

Vorobyov* then took the floor. He said that the instrument he was using recorded high radiation fields. In the 250-roentgen range it went off the scale in the vicinity of the pile of rubble, the turbine hall, the central hall, and other places in and around No. 4 unit.

In conclusion, he recommended an immediate evacuation,

---

*He was the same director of civil defense at the plant who, in the first two hours after the accident, used the one and only 250-roentgen radiometer to detect the dangerous radiation levels, and then reported to Bryukhanov. The reader is familiar with Bryukhanov's reaction. Also, that same night Vorobyov alerted the civil defense headquarters of the Ukraine and deserves to be commended for it.

whereupon Bryukhanov tried to silence his subordinate, warning him not to aggravate matters.

The meeting was then addressed by V. D. Turovsky, representative of the Ministry of Health: "Anatoly Ivanovich, we must order an immediate evacuation, after what we saw in the medical center. Those patients are in very serious condition. As far as we can tell after an initial examination, they must have been exposed to radiation between three and five times higher than the lethal dose. They're suffering from extremely severe external and internal irradiation. Their skin is the brown color characteristic of a nuclear tan. There can be no doubt that radioactivity has been disseminated far from the reactor unit."

"What if you're mistaken?" asked Mayorets, as calm and composed as ever, but clearly straining to keep his temper. "Let's find out the situation and take the correct decision, but I am opposed to evacuation. The danger is clearly being exaggerated."

The meeting was suspended. During the recess the minister and Shasharin went into the corridor for a smoke.

Testimony of B. YA. PRUSHINSKY, chief engineer of Soyuzatomenergo (Department of Nuclear Energy):

When Kostya Polushkin and I returned to the Party Committee of the town of Pripyat, Shasharin and Mayorets were standing in the corridor smoking. We went up to them and, in fact, reported the findings of our aerial inspection of No. 4 unit to the minister right then and there.

Shasharin noticed us first and invited us to describe what we had seen from the helicopter. He was puffing away strenuously on his cigarette, in the midst of clouds of smoke. He was a puny man who, when attending party meetings, had seemed rather like a toy next to a rough, heavyweight figure like the deputy minister A. N. Semyonov. But now he seemed even paler and more shrunken, his usually smooth brown hair all disheveled. As he stared at us unblinkingly from behind the huge lenses of his imported glasses, his light blue eyes had a haunted look about them. Actually, all of us were dead tired and had a slightly hunted look, with the possible excep-

tion of Mayorets. He was his usual dapper self, his hair parted neatly on his pink skull, his round face as expressionless as ever. Perhaps he simply failed to understand. That's probably what it was.

I briskly began telling them how Polushkin and I had inspected No. 4 unit from the air. From an altitude of 820 feet (250 m) we had seen that the reactor was destroyed. In other words, what was destroyed was the main circulation pump rooms, the drum-separator rooms, and the central hall. The upper biological shield of the reactor was now a bright cherry-red from the extreme heat and was lying at an angle over the reactor vault. On it we could quite easily see the wreckage of the system for monitoring the integrity of the fuel channels and of the protection and control system. Graphite and fragments of fuel assemblies were scattered about everywhere: on the roofs of V block, the turbine hall, the de-aerator, on the asphalt around the unit, and even among the 330- to 750-kilovolt switching station. It was safe to say that the reactor had been destroyed. The cooling was having no effect. Polushkin agreed that the reactor was utterly finished. When Mayorets asked us what should be done, I replied, "God knows. Right now I can't say. There's graphite burning in the reactor. That has to be extinguished, before anything else. But how, and with what? We've got to think."

We all went into Gamanyuk's office. The government commission continued its work. Shasharin put forward the idea of setting up working groups. When the question of restoring the damaged unit was mentioned, Konviz, the representative of the general designer, said, without being invited to speak, that he thought entombment would be more appropriate. Mayorets cut him off, urging him to keep to the subject under discussion, and then announced that the groups referred to should have their recommendations ready in an hour for Shcherbina, who was expected an hour or two later.

Testimony of G. A. SHASHARIN:

Together with Marin and Sidorenko—the deputy chairman of the Committee on Operational Safety in the Nuclear Power Industry, and the corresponding member of the Academy of Sciences—I then went up in the helicopter to a point 800–1,000 feet (250–300

m) above the damaged reactor unit. The pilot had a dosimeter; rather, I think it was a radiometer. At that altitude, the radiation level was about 300 roentgens per hour. The upper biological shield was now bright yellow from the heat, instead of the cherry-red reported by Prushinsky, which meant that the temperature had increased. The biological shield was not tipped at such a pro- nounced angle, on top of the reactor vault, as it was later on, when bags of sand were dropped. The added weight caused it to shift.

Now I became absolutely convinced that the reactor had been destroyed. Sidorenko suggested dropping about 40 tons of lead to reduce the radiation. I was categorically opposed to this idea, as such a weight, especially coming from a height of 650 feet (200 m), would have had an enormous impact, punching a hole right through to the suppression pool, and driving the entire molten core down into the water below. Then everybody would have had to flee for their lives.

When Shcherbina arrived, I went to see him privately before the meeting, told him the situation, and said that the town should be evacuated immediately. He calmly replied that this could cause panic, and that panic was even worse than radiation.

About that time, approximately 7 P.M., the plant completely ran out of water reserves; and the pumps, which the electricians had started with such difficulty and at the cost of such immense radiation exposure, stopped functioning. As we have seen, all the water had found its way not into the reactor, but into the underground com- partments, where it flooded the electrical installations of all the plant's reactor units. Radioactivity was increasing rapidly every- where, and the destroyed reactor continued to spew forth millions of curies of radioactivity from its red-hot jaws. The air was filled with the entire range of radioactive isotopes, including plutonium, americium, and curium. All of these isotopes were absorbed into the bodies of people working at the plant and of residents of Pripyat. Throughout 26 and 27 April, right up until the evacuation, radionu- clides continued to accumulate in people's organisms; moreover, they were exposed to external gamma and beta radiation.

AT THE PRIPYAT MEDICAL CENTER

The first group of victims, as we already know, was brought to the medical center 30 or 40 minutes after the explosion. We should not forget the special gravity of the highly unusual situation resulting from the nuclear disaster at Chernobyl, in which the effect of radiation on the human organism proved rather complex: severe external and internal radiation was further complicated by heat burns and moistening of the skin layers. The actual degree of injury and exposure could not be promptly determined, because the plant's radiation safety service had not supplied the doctors with accurate data about the radiation fields. As we have seen, the radiometers on hand at the plant showed a radiation intensity of 3 to 5 roentgens per hour. At the same time, the more accurate information provided by Vorobyov, the head of the plant's civil defense unit, was ignored. The milder version supplied by the radiation safety service was not a sufficient warning to the doctors in the medical center, who were in any case inadequately prepared for such eventualities.

In fact, the only clues to the gravity of the injuries were provided by the clinical signs evident in the patients when they were brought in: pronounced erythema (nuclear tan), edema, burns, nausea, vomiting, weakness, and certain conditions indicating shock.

Moreover, the medical center serving the Chernobyl nuclear power station was not equipped with the necessary broad-range radiometric apparatus, capable of promptly identifying the nature and degree of external and internal irradiation. It is quite evident that the doctors at the medical center were not prepared, from an organizational point of view, for patients such as these. In such cases, each individual should be immediately classified according to the progression of acute radiation syndrome, on the basis of certain characteristic initial reactions, the differences between which vitally affect the choice of treatment. However, such classifications were not made. The probable outcome of the illness was regarded as the basic criterion:

1. Recovery impossible or unlikely.
2. Recovery possible with the use of modern treatment.

3. Recovery probable.
4. Recovery guaranteed.

Such a classification is particularly important when, as a result of an accident, large numbers of people have been irradiated, and it may prove necessary to promptly identify those whose lives might be saved by timely medical assistance. Such assistance, in other words, should be provided to persons in categories 2 and 3, as their fate substantially depends on timely treatment. Here it is particularly important to know when the exposure started, how long it lasted, and whether the skin was dry or wet, as radionuclides penetrate wet skin more intensively, especially if it has been injured by burns or wounds.

We know that virtually the whole of Akimov's shift lacked respirators and protective tablets of potassium iodide and Pentocin and worked without proper dosimetric monitoring.

None of the patients sent to the medical center were classified according to the progression of their acute radiation sickness, and all mixed freely with one another. Decontamination of the skin was inadequate, consisting merely of a shower, which had little or no effect because of the dissemination of radionuclides with accumulation in the granular layer beneath the epidermis.

Priority was given to the treatment of those patients from group 1, with severe initial reactions, who were promptly connected to an intravenous line; and of those with severe heat burns (Shashenok, Kurguz, and the firefighters).

By the time a specialized team flew in from Moscow, consisting of physicists, radiologists, and blood specialists, fourteen hours had elapsed since the accident. Blood tests were done as many as three times; medical records were prepared, including clinical signs since the accident, the nature of a patient's complaints, and the number and formula of the lymphocytes.

Testimony of V. G. SMAGIN, shift foreman of No. 4 unit, who relieved Akimov:

About 2 P.M., after I had started vomiting, with dizziness, headaches, and fainting spells, I left the control room, washed and changed in the personnel air lock, and went to the infirmary in No. 1 administrative building. The doctors and nurses there tried to write down where I had been and the level of background radiation I had been exposed to. But we knew nothing. All we knew was that at 1,000 microroentgens per second the instrument went off the scale. Where I had been? How was I supposed to remember where I had been? You'd have to describe to them the entire layout of the plant. On top of that I felt really lousy all the time. Then five of us were taken by ambulance and brought to the medical center in Pripyat.

We were taken into the reception room, and each of us was checked for radioactivity with a special measuring device. All of us were radioactive. We washed again. We were still radioactive. Then we went upstairs to the first floor, to the doctors. Lyudmila Ivanovna Prilepskaya took charge of me as soon as she saw me. Her husband was also a shift foreman, and our two families had been very friendly. At that point all of us started to vomit. We saw a bucket or pot and would start throwing up right into it.

Prilepskaya noted down the details of my case and asked what part of the unit I had been in and what the radiation fields were like there. She just could not get it into her head that there were radiation fields and contamination everywhere. There was not a single clean spot left. The entire power plant was one colossal radiation field. She tried to determine the amount of my exposure. In the intervals when I wasn't vomiting, I tried to tell her as best I could. I told her that none of us knew what the field really was. All we knew was that the instrument went off the scale at 1,000 microroentgens per second. I felt really ill, very weak, nauseated, and dizzy.

I was taken into a room, put in an empty bed, and immediately hooked up to an intravenous line for between two and a half and three hours. I had three bottles fed into me, two of them containing a clear fluid, and in the third it was yellowish.

Two hours later I began to feel better. When the IV was finished, I got out of bed and went to look for a cigarette. There were two others in the room. In one bed there was a militiaman who kept on saying that he wanted to go home, that his wife and children were worried, they didn't know where he was, and he didn't know how they were. I told him to stay in bed; he had been exposed to radiation and now needed treatment for it.

In the other bed there was a young adjuster from the firm handling the start-up of the Chernobyl plant. When he learned that Volodya Shashenok had died that morning, probably at 6 A.M., he started shouting, asking why he hadn't been told. He was hysterical. He must also have been really scared, figuring that since Shashenok was dead he might die, too. He was yelling at the top of his voice, "Why was his death kept a secret? Why wasn't I told about it?" Then he calmed down, but a severe bout of hiccups left him exhausted.

The medical center was contaminated, as the instrument showed. Some women brought over specially from Yuzhatomenergomontazh were washing the corridor and the rooms nonstop. A dosimetrist was walking around measuring everything; I could hear him muttering that, despite all the washing, radiation levels were still high.

It sounded as if he was not satisfied with the way the women were working, but in fact they were really doing their best and were not to blame. The windows were wide open; it was stiflingly hot outside; and there was radioactivity, including gamma radiation, in the air. That's why the readings on the instrument were inaccurate—though actually the contamination they showed was genuine—because it all came floating in through the window and settled over everything.

I heard my name called through the open window. I looked out and saw Seryozha Kamyshny, the reactor foreman from my shift, in the street below. When he asked how I was, I answered that I needed a smoke. I let down a string and hoisted some cigarettes back up. Then I said, "What about you, Seryozha, what are you doing out there? You picked up some radiation, too. Come on in here." He replied that he was feeling fine, and, taking a bottle of vodka out of his pocket, said, "See, I've just been decontaminating myself. Like some?" I refused, saying that I had already had an IV.

I looked into the next room where Lenya Toptunov was in bed. He had turned completely brown. His mouth, lips, and tongue were all terribly swollen, and he could hardly talk. Everyone was wonder-

ing the same thing: What had caused the explosion? I asked him about the reactivity margin. With the greatest effort he managed to say that the Skala computer showed 18 rods, but that it could have been wrong. Machines occasionally made mistakes.

Volodya Shashenok had died from burns and radiation at about 6 A.M. I was told he had already been buried in a village cemetery. But Aleksandr Lelechenko, deputy chief electrical engineer, felt so good after his IV that he left the medical center and went back to work at the unit. The next time he was brought back in was in Kiev, and in critical condition. He died there in agony. His overall dose came to 2,500 roentgens. Nothing could be done for him—intensive treatment and bone marrow transplants were no help.

After the IV many of them felt better. In the corridor I met Proskuryakov and Kudryavtsev, both of whom had their arms tight against their chests. That was how they had crossed them to protect themselves from the radiation from the reactor in the central hall, and that's the way they stayed; it was too painful to straighten them out. Their swollen faces and hands had turned dark brown. Both complained of intense pain in the skin of their faces and hands. They were unable to spend any longer talking, and I didn't want to upset them any more.

But Valera Perevozchenko did not get up after his IV. On various parts of his body the skin had burst and was hanging off in strips. His face and hands were severely swollen and covered with scabs, which broke open at the slightest movement. And he was in great pain. He complained that his whole body had turned into one massive pain.

Petya Palamarchuk, who had carried Volodya Shashenok out of the nuclear inferno, was in similar condition.

Of course, the doctors worked hard to help their patients, but there was only a limited amount they could do. They had become irradiated themselves. The air in the medical center was radioactive. And the severely ill patients were themselves emitting radiation, as they had absorbed radionuclides through their skin.

The truth is, nothing like this had ever happened anywhere in the world. We were the first after Hiroshima and Nagasaki, though that was nothing to be proud of.

All those who were feeling a little better assembled in the smok-

ing room. They were all thinking the same thing: Why had the explosion occurred? Sasha Akimov was there, looking rather pathetic and very dark brown. Anatoly Stepanovich Dyatlov came in smoking, looking his usual thoughtful self. Someone asked him what dose he had received, to which he replied that he thought it must have been around 40 roentgens, and that he was going to survive.

He was wrong about the 40 roentgens—exactly ten times too low, in fact. In No. 6 clinic in Moscow, they found that he had been exposed to 400 roentgens, and that he had third-degree radiation sickness. His legs had been heavily irradiated when he walked through the graphite and fuel around the damaged unit.

Why had it all happened? Everything was proceeding normally. Everything had been done correctly, and the reactor was in a relatively stable mode, when suddenly, in a matter of seconds, everything had fallen apart. That's what all the operators thought.

The general view was that only Toptunov, Akimov, and Dyatlov could answer questions. The catch was, however, that they could not answer this question. Many people were beginning to think it might have been sabotage, because when something cannot be explained—well, you start thinking of the most outlandish theories.

In answer to my question, Akimov said just one thing, "We did everything properly. I don't understand why it happened." He was angry and bitter.

At the time many people failed to understand. We hadn't yet grasped the full gravity of the disaster that had overtaken us. Dyatlov was also convinced he had done the right thing.

That evening a team of doctors arrived from No. 6 clinic in Moscow and walked around the rooms examining us. A bearded doctor—I believe, Dr. Georgi Dmitriyevich Selidovkin—chose the first batch of twenty-eight patients to be sent to Moscow immediately. Dispensing with tests, he based his choice on their nuclear tan. Almost all twenty-eight died.

We could see the damaged reactor building clearly from the window of the medical center. At night the graphite flared up violently, wrapping a huge spiral of flames around the ventilation stack. It was a terrible, painful sight.

Sasha Esaulov, the deputy chairman of the local Party Executive

Committee arranged the departure of the first group of patients. Kurguz and Palamarchuk were taken in an ambulance, and the other twenty-six traveled in a red Ikarus bus to Borispol airport. They took off for Moscow at 3 A.M.

The remaining patients whose condition was less serious, including me, were sent to No. 6 clinic on 27 April. After a tearful and noisy sendoff, about a hundred of us left Pripyat around midday in three Ikarus buses, still wearing our striped hospital garments.

In No. 6 clinic they found that I had received a dose of 280 rads.

FIRST OFFICIAL ACTIONS

About 9 P.M. on 26 April 1986, Boris Yevdokimovich Shcherbina, deputy chairman of the Council of Ministers of the USSR, arrived in Pripyat. As the first chairman of the government commission set up to eliminate the consequences of the nuclear disaster in Chernobyl, he had a truly historic role to perform. In my opinion, however, his entire management of the power industry, through the incompetent Mayorets, had hastened the onset of disaster.

This rather puny man, of medium build, now a little paler than usual, his tight-lipped face showing its age, his thin cheeks deeply marked with the lines of authority, was calm, collected, and focused. He still did not realize that the air around him—in the street and inside the room—was saturated with radioactivity, and emitting gamma and beta rays which penetrated whomever happened to be in their way—ordinary mortals, Shcherbina, or the devil himself. As for ordinary mortals, there were about 48,000 of them in the town that night, including senior citizens, women, and children. But something of the same indifference was to be found in Shcherbina, who alone was empowered to decide whether to evacuate, and whether to classify what had happened as a nuclear disaster.

His behavior was typical of the man. Initially he seemed quiet, modest, and even a little apathetic. This small and frail individual evidently savored the colossal power he wielded, which was so hard to keep in check; he thought of himself as a godlike figure, with the ability to punish, or forgive, as he pleased. However, Shcherbina was just human, as his later behavior showed. At first he appeared outwardly composed, but gradually a storm was gathering strength within him, and by the time he had understood the disaster and decided on a policy for coping with it, he burst forth in a frenzy of impatient energy, driving everyone relentlessly to work harder and faster.

In Chernobyl a cosmic tragedy had taken place; and the cosmos can be handled not by brute force alone, but by the force of reason, which is in itself a living and more powerful cosmic force.

Mayorets first took the floor to sum up the work of the working groups. He was obliged to acknowledge that No. 4 reactor unit had been destroyed and the reactor with it. He then outlined the various methods for enclosing, or burying, the unit, which included depositing more than 250,000 cubic yards (200,000 m³) of concrete in the destroyed reactor building. Metal caissons would have to be erected around the unit and then filled with concrete. He did not know what needed to be done with the reactor, which was extremely hot. Evacuation had to be considered, but he was hesitant about it, feeling that if the reactor was totally choked off, the radioactivity would decline or disappear altogether.

"Don't rush into any evacuation," said Shcherbina quietly, but obviously straining to suppress his innermost feelings of impotent rage. He so wanted to avoid an evacuation! Mayorets had made such a good start in the new ministry; the utilization rate for installed capacity had gone up; and frequency in the power system had become stabilized. And now this . . .

Mayorets was followed by a number of speakers: Shasharin, Prushinsky, General Berdov, Gamanyuk, Vorobyov, Lieutenant General Pikalov (chief of the chemical warfare forces), Kuklin and Konviz on behalf of the designers, and Fomin and Bryukhanov for the management of the Chernobyl plant. Once he had heard all the speakers, Shcherbina invited everyone to put their heads together: "What we need now, comrades, is a brainstorming session. I refuse to believe that some reactor or other cannot be extinguished. After all, huge firestorms from burning gas wells have been extinguished, so why not a reactor?"

The brainstorming session then began, with everybody suggesting whatever came into his head. Herein lies the merit of brainstorming sessions, as even far-fetched, nonsensical, or heretical ideas might suddenly provide a clue leading to a sensible course of action. Among other things, it was suggested that an enormous tank filled with water should be hoisted by helicopter and dropped into the reactor; that a kind of nuclear Trojan horse should be made, consisting of a huge hollow concrete cube which would enable people to move close to the reactor and then, when in position, drop something into it. Someone sensibly inquired how the concrete device, or Trojan horse, would be moved, thus raising the question of suitable wheels and motors. The idea was dropped.

Shcherbina made a suggestion himself. It was that fire boats should be moved up the intake channel, right next to the unit, and pour water over the burning reactor from there. One of the physicists explained that nuclear fire cannot be extinguished with water, as it would merely increase the radiation. The water would evaporate, mix with nuclear fuel, and then contaminate everything around. The idea of the fire boats was also dropped.

Eventually someone remembered that sand harmlessly extinguishes fire, including the nuclear variety. At this point, it became evident that aircraft were going to be indispensable, so the people in charge of helicopters were summoned from Kiev.

Major General Nikolai Timofeyevich Antoshkin, deputy commander of the air force of the Kiev military district, was already on his way to Chernobyl, having received an order from the district on the evening of 26 April. It read as follows: "Go immediately to Pripyat. The decision has been taken to dump sand on the damaged nuclear reactor. The reactor is 100 feet (30 m) high. Helicopters are clearly the only technology suitable for this job. In Pripyat act according to the circumstances. Keep in constant touch with us."

The military helicopters were stationed far from Pripyat and Chernobyl and would have to be moved closer.

While General Antoshkin was on his way, the government commission decided to order an evacuation—something the representatives of civil defense and the doctors from the Ministry of Health had been insisting on adamantly.

Vorobyov, the deputy minister of health, made an impassioned plea for immediate evacuation: "The air is full of plutonium, cesium, and strontium. The condition of the patients in the medical center could have been caused only by very high radiation fields. Everyone's thyroid glands, including those of children, are packed with radioactive iodine. Nobody has been taken potassium iodide as a preventive measure. It's really shocking!"

Shcherbina interrupted him, "We're evacuating the town on the morning of 27 April. I want all eleven hundred buses lined up at night on the highway between Pripyat and Chernobyl. You, General Berdov, will kindly assign guards to each house and let no one out on the street. Civil defense will convey the necessary information, including the detailed timing of the evacuation, over the radio. Distribute potassium iodide tablets to each apartment. Use the members of the Communist Youth League for this purpose. And now I'm going to take a look at the reactor from the air, with Shasharin and Legasov. It will be easier to see at night."

Shcherbina, Shasharin, and Legasov flew in a civil defense helicopter into the radioactive night sky over Pripyat and hovered above the reactor. Shcherbina looked through binoculars at the reactor, now bright yellow from the extreme heat. Against that background he could clearly see the dark smoke and tongues of flame. And a glimmering blue light, not unlike starlight, shone from deep gashes to both right and left, deep within the bowels of the destroyed core. It seemed as if some immensely powerful hand was pumping invisible bellows, fanning that gigantic nuclear furnace, 65 feet (20 m) in diameter. Shcherbina was impressed as he gazed into that fiery nuclear monster, which unquestionably had far greater power than the deputy chairman of the Council of Ministers of the USSR. In fact, it was so much greater that it had already demolished the careers of many high-ranking officials and could well remove him, Shcherbina, from his post. This was clearly an adversary to be reckoned with.

"That's quite some fire!" said Shcherbina quietly, as if talking to himself. "And how much sand has to be dumped into that crater?"

"When fully assembled and loaded with fuel, the reactor weighs 10,000 tons," Shasharin replied. "If half the graphite and fuel have been ejected, that would be about 1,000 tons, and would have left a hole as much as 13 feet (4 m) deep and about 65 feet (20 m) wide.

Sand is denser than graphite. I think between 3,000 and 4,000 tons of sand will have to be dumped."

"There'll be plenty of work for the helicopter pilots," said Shcherbina. "What is the radioactivity level at 820 feet (250 m)?"

"Three hundred roentgens per hour. But the impact of the materials landing in the reactor will kick up nuclear dust, and radioactivity at that altitude will increase substantially. We're going to have to bomb it from a lower altitude."

The helicopter descended from the crater. Shcherbina was relatively calm; but his calm was attributable not only to the composure normal in one of his rank but also, and to a great extent, to his ignorance of nuclear matters as well as to the uncertainty of the situation. Within a few hours, after the first decisions had been taken, he would start shouting at his subordinates at the top of his voice, hurrying them along and accusing them of slowness and every crime under the sun.

4

27 APRIL 1986

Testimony of Colonel V. Filatov:

It was already well after midnight on 27 April when Air Force Major General Antoshkin entered the Party Committee building in Pripyat. On the way into Pripyat, he had already been struck by the fact that the lights were on in all office windows. Instead of sleeping, the town was humming, like a disturbed beehive. The Party Committee building was crammed with people. He reported at once to Shcherbina.

Shcherbina said, "Everything depends on you and your helicopter pilots now, general. The crater has to be sealed off tightly with sand, from above. From now on, there is no other way of getting into the reactor. Only your helicopter pilots can do it."

When General Antoshkin asked when he was supposed to start operations, Shcherbina looked at him in amazement and told him he would have to move into action immediately.

"That's impossible, Boris Yevdokimovich," protested the speaker. "Our helicopters still have to move to their new base. We

have to find a landing area, a flight control center. It'll have to wait till dawn."

Shcherbina agreed, but emphasized that operations would start at the crack of dawn, not a moment later. He told the general he hoped he understood what was involved, and instructed him to take charge of the conduct of this operation.

General Antoshkin, feeling slightly overwhelmed, frantically tried to think of a plan. Where was he to get all that sand? And all those bags? Who would load them onto the helicopters? What would be their flight path to No. 4 reactor? From what altitude should the bags were dumped? What were the radiation levels? Was it at all possible to send pilots over the crater? What if they were suddenly taken ill in midair? Traffic control would be needed. Who would do it, how, and from where? What bags of sand? Was he supposed to create them all from nothing?

He then devised a plan with the following components: bags of sand—helicopters, dumping of bags of sand; distance from landing pad to crater; landing pad—place of deployment; reactor—radiation—decontamination of crews and equipment.

Antoshkin then remembered that, as he was driving from Kiev to Pripyat, he had seen an endless stream of buses and private cars going the other way, all of them crammed with people as in rush hour. The thought of evacuation suddenly occurred to him. This was a spontaneous evacuation, some of the inhabitants having begun to leave the radioactive town on their own initiative as early as the daytime hours and the evening of 26 April.

Antoshkin pondered where to station his helicopters. And all of a sudden the square in front of the Party Committee offices in Pripyat came to mind. It occurred to him that this was the perfect place.

He told Shcherbina what he intended to do. After some hesitation, as the engine noise would interfere with the work of the government commission, Shcherbina agreed.

Oblivious to the varying radiation levels around him, Antoshkin drove straight to the damaged reactor building to examine the flight paths the helicopters would have to take. And he did so without any means of protection. In its confusion, the plant management had failed to supply the visitors with any protection. They all went about their business in the clothes they had arrived in. By the end of 24 hours their hair and clothes were heavily radioactive.

EVACUATION

Long after midnight on 27 April, Major General Antoshkin sent for the first pair of helicopters over his personal two-way radio. However, in the circumstances, without a ground controller, they were unable to land, so Antoshkin went up to the roof of the ten-story Pripyat Hotel with his two-way radio and acted as a flight controller. He could see the heavily damaged No. 4 reactor building, with its corona of flames above the reactor, as if it were in the palm of his hand. Off to the right, beyond the Yanov station and the overpass lay the road to Chernobyl, on which there was an endless and colorful row of empty buses, only dimly visible in the early morning mist—red, green, blue, yellow, all motionless, waiting for the order to move.

The eleven hundred buses were parked for 12 miles along the road from Pripyat to Chernobyl. It was a most depressing sight. As their chillingly empty windows glinted in the rays of the rising sun, they seemed to be stark symbols of the death of this ancient and once pure land.

At 1:30 P.M., this column started to move: it crawled across the overpass and spread out around the town, with one bus stopping outside each white apartment building. And then as it left Pripyat, taking its inhabitants away forever, its wheels also carried away vast amounts of radioactivity, thus contaminating the roads of villages and towns along its route.

The tires of these vehicles should, of course, have been changed before the convoy left the 6-mile (10-km) zone; but no one thought of doing so. For a long time thereafter, radiation levels on the asphalt of Kiev varied between 10 and 30 millibers per hour, and the roads had to be washed down regularly for months.

The final decisions on the whole question of evacuation were taken in the middle of the night. The prevailing view, however, was that the evacuation would last only two or three days. The scientists meeting in the Party Committee building in Pripyat assumed that the radiation would decrease after the reactor had been sealed off with sand and clay. Admittedly, science itself was still somewhat vague about these issues, but nonetheless the idea that the radiation would be short-lived carried the day. The public was therefore told to wear light clothes, to take food and money for three days, put the rest of their clothes away in drawers, turn off the electricity and gas, and lock their doors. Their apartments would be protected by the militia.

If the members of the government commission had known the magnitude of the background radiation, they would have reached quite a different decision. Many residents could have carried with them their main items of clothing, packed in plastic bags. Like normal dust, radioactive dust continued to enter apartments through cracks around the doors and windows; within a week, the radioactivity of objects left in the apartments had risen to 1 roentgen per hour.

Many women left wearing thin bathrobes and dresses and taking with them, especially on their hair, millions of nuclear particles.

Testimony of VLADIMIR N. KOLAYEVICH SHISHKIN, deputy director of Soyuzelektromontazh, in the Ministry of Energy of the USSR:

The original idea was to evacuate the town early in the morning. Shasharin, the Ministry of Health, Vorobyov, Turovsky, and the representatives of civil defense all strongly favored this idea.

The scientists had nothing to say about the evacuation. In fact, it seemed to me in general that they tended to downplay the hazards. They were extraordinarily vague and uncertain about what to do with the reactor. Dumping sand onto it was viewed as a preventive means of combating the fire in the reactor.

Testimony of B. YA. PRUSHINSKY, chief engineer of Soyuzatome-nergo (Department of Nuclear Energy):

On 4 May, I flew over the reactor in a helicopter with Academician Velikhov. After studying the destroyed reactor building from the air, Velikhov admitted that he did not know how to bring the reactor under control. He sounded quite worried. And this was after

5,000 tons of assorted materials had been dumped into the gaping hole left by the nuclear blast.

Testimony of V. N. SHISHKIN:

By 3 A.M., it became evident that a morning evacuation would be impossible, both organizationally and technically. The public had to be warned. It was then decided to hold a meeting in the morning of representatives of all the enterprises and organizations in the town, so as to explain the evacuation in detail.

None of the members of the commission wore respirators, and no one had issued potassium iodide tablets. In fact, no one asked for them. Here again, science had failed to rise to the occasion. Bryukhanov and the local authorities were well out of their depth, and Shcherbina and many of the other members of the commission, including me, knew nothing about dosimetry and nuclear physics.

I later learned that the radioactivity in the room we were in reached 100 millibers per hour, or 3 roentgens per 24 hours if one stayed indoors, while outside the level was as high as 1 roentgen per hour, or 24 roentgens per 24 hours. And that was only external radiation. Iodine-131 accumulated in the thyroid gland much faster; and as the dosimetrists explained to me later, by midday on 27 April the radiation being emitted by the thyroid gland of many people was as much as 50 roentgens per hour. The ratio of thyroid to whole body radiation is 1:2. This meant that people were receiving another 25 roentgens from their own thyroids, in addition to the dose received from external sources. The total dose received by each resident of Pripyat and each member of the government commission by 2 P.M. on 27 April was, on average, in the range of 40 to 50 rads.

At 3:30 A.M., I felt my knees buckling under me from extreme fatigue—nuclear fatigue, as it later turned out—and I went to take a nap.

On the morning of 27 April, I woke up around 6:30 and went out onto the balcony for a smoke. From the next balcony of the Pripyat Hotel, Shcherbina was intently studying the damaged reactor building through a telescope.

Around ten o'clock all the representatives of the town's enterprises and organizations assembled to have the situation explained

to them and to be given instructions for the evacuation, now scheduled for 2 P.M. The main objectives were preventing people from
going outside, distributing potassium iodide tablets, and washing
down apartments and city streets. Dosimeters were not handed out,
as they were in short supply; the ones in the damaged reactor
building were contaminated.

All the members of the government commission had lunch and
dinner on 26 April, and breakfast, lunch, and dinner on 27 April,
without any precautions in the restaurant of the Pripyat Hotel,
where they ingested radionuclides together with their food. It was
not until the evening of 27 April that, on the insistence of civil
defense, dry rations were issued, consisting of sausage, cucumbers,
tomatoes, melted cheese, coffee, tea, and water. Everyone received
their rations, except for Mayorets, Shcherbina, and Marin who, as
usual, were waiting to be served. And by the time they went along
for their portions, everything was gone. This gave rise to
great merriment.

Around midday on 27 April, the members of the government
commission were all feeling more or less the same, with nuclear-
induced fatigue (which becomes noticeable earlier and is more acute
than the fatigue produced by the same amount of work in normal
conditions), tickling in the throat, dryness, coughing, headache, and
itching skin.

On 27 April, during daylight hours, an hourly dosimetric survey
of Pripyat was carried out. Samples of asphalt, air, and roadside dust
were collected. An analysis showed that 50 percent of the radioactive particles consisted of iodine-131. Radioactivity at ground level
on the streets was as much as 50 roentgens per hour, and 6.5 feet
(2 m) above the ground about 1 roentgen per hour.

Testimony of MIKHAIL STEPANOVICH TSVIRKO, director of the All-
Union Nuclear Power Construction and Assembly Enterprise,
Soyuzatomenergostroy:

On the evening of 27 April, all the cooks vanished. The water no
longer came out of the faucets. There was nowhere to wash your
hands. They brought us pieces of bread in cardboard boxes, cucumbers in another box and jam in a third, and other things as well.

*Taking no chances, I picked up the bread, took a bite, and then
threw away the piece I had been holding in my hand. I later realized
I had been wasting my time, because the piece of bread I actually
ate was just as contaminated as the one I was holding in my hand.
Everything was severely contaminated.*

Testimony of Irina Petrovna Tsechelskaya, secretary at the Pripyat cement-mixing plant:

*We were told that the evacuation was for three days only, and
that there was no need to take anything, so I left wearing my
bathrobe. All I took with me was my internal passport and a little
money, which soon ran out. After the three days were up, they did
not let us back in. I got as far as Lvov and had no money. If I had
known, I would have taken my savings bank passbook; but I left
everything behind. When I showed my Pripyat residency permit as
proof, no one took any notice. They couldn't have cared less. I
applied for an allowance, but they wouldn't give me one. I wrote
a letter to the minister of energy, Mayorets. I don't know, but I
imagine my bathrobe and everything on me was highly contaminated. Nobody measured me.*

Mayorets noted on Tsechelskaya's letter: "Comrade Tsechelskaya
should apply to any organization of the Ministry of Energy. She will
be given 250 roubles." The note, however, was dated 10 July 1986,
whereas these events took place on 27 April.

Testimony of G. N. Petrov, former head of the equipment section
at the Pripyat branch of Yuzhatomenergomontazh:

*On the morning of 27 April, we heard over the radio that we were
not to leave our apartments. Health workers went around distributing potassium iodide tablets. A militiaman, not wearing a respirator,
was posted outside each apartment building. As we later discovered,
there was about 1 roentgen per hour on the streets and radionuclides
in the air.*

*Not everybody obeyed instructions. It was a warm, sunny day,
and a Sunday, moreover. But people were coughing, with dry
throats, headaches, and a metallic taste in their mouths. Some went*

over to the medical center to get measured. Their thyroids were checked with an measuring device which went off the scale at 5 roentgens per hour. Those were all the instruments they had, so the actual radiation levels were uncertain. People were getting very nervous, but then they soon forgot. They were extremely agitated.

Testimony of LYUDMILA ALEKSANDROVNA KHARITONOVA, senior engineer in the construction department of the Chernobyl nuclear power station:

As early as the afternoon of 26 April, some people, particularly schoolchildren, were warned not to go outdoors. But hardly anyone took any notice. By evening it had become obvious that there really was good reason to be alarmed. Everyone was discussing their apprehensions with friends and neighbors. I didn't see this myself, but I understand that many people, especially men, decontaminated themselves with liquor. There were drunks around the workers' settlements even in normal times, but now, with the nuclear disaster, there was an added inducement. The fact is that liquor was the only thing people had for decontamination. Pripyat was remarkably lively, with swarms of people milling about, as if a gigantic carnival was about to take place. Of course, the May Day celebrations were coming up; but even so, the hyperagitated state of the people was quite remarkable.

Testimony of LYUBOV NIKOLAYEVNA AKIMOVA, the wife of the shift foreman of No. 4 unit:

On the morning of 27 April, we heard over the radio that we were not allowed to go outdoors or anywhere near the window. High school girls brought around potassium iodide tablets. At midday it was made clear that an evacuation would be taking place, but that it wouldn't be for long, only for two or three days. We were urged to stay calm and not to take many things. The children kept going to the window to see what was happening out on the street, and I kept pulling them away. We were scared. I looked through the window and realized that not everybody was following instructions. A woman who lived next door was sitting on a bench outside the apartment building, knitting, while her two-year-old son p!ayed in the sand. As we were told later, the air we were breathing was emitting gamma and beta rays, and saturated with long-lived radionuclides—all of which were accumulating in our bodies, especially radioactive iodine in the thyroid gland, which posed a special threat to children. I had a constant headache and a terrible hacking cough.

On the whole, though, everyone led normal lives, preparing breakfast, lunch, and dinner, going to the shops all day and into the evening of the 26th, and also on the morning of the 27th. People were going to each other's apartments.

The trouble is, the food we were eating was also contaminated with radiation. I was really worried about my husband's condition; his skin had turned dark brown, he was extremely agitated with a feverish glint in his eyes.

Testimony of G. N. Petrov:

Buses rolled up outside each apartment building at exactly 2 P.M. We had been warned already over the radio to dress lightly and take as little as possible with us, as we would be returning within three days. I found myself wondering what this really meant, because if we had taken lots of things even 5,000 buses would not have been enough.

Most people did what they were told and didn't even take spare money with them. Actually our people behaved very well, joking, comforting each other, and calming the children by telling them they were going to see grandma, to the movies, or to the circus. The older children looked pale and sad and said nothing. Besides radiation, there was also alarm and contrived jollity in the air. Many people went downstairs early and waited outside, together with the children, though they were continually being told to go back in. When the time came, we went straight from the entrance and boarded the buses. Some who got confused started rushing from one bus to the next, thus picking up extra radiation. In this way, on that typical "peaceful" day we absorbed more than our fair share, inside and out.

We were driven to Ivankov, 37 miles from Pripyat, and then to various villages. We were not always given a warm reception. One *resident would not let us into his enormous brick house, not because of the radiation (which he didn't understand anyway and it was no good explaining it to him) but out of sheer selfishness. He said he had not built the place so as to let strangers in.*

Many of those who were deposited in Ivankov went farther, toward Kiev, on foot; some of them hitchhiked, with no idea of what they expected to find. Some time later, a helicopter pilot I know told me that he had seen, from the air, enormous crowds of lightly clad people, women and children, and old people walking along the road, and on the side of the road, in the direction of Kiev. They had already reached Irpeni and Brovarov. Cars were stuck in the midst of these crowds, as if they were among vast herds of cattle being driven to pasture. In the movies you often see this kind of scene in Central Asia, and that's exactly what came to mind, though the comparison was not a very nice one. And the crowds of people kept on walking, endlessly.

When the time came to say goodbye to their pets, there were some distressing scenes. The cats, with their tails straight up, stared at their masters with an imploring look, miaowing pathetically; and dogs of many different breeds were whining plaintively, trying to force their way into the buses, yelping frantically and growling when they were pulled away. However fond the children were of their pets, there could obviously be no question of taking cats and dogs, as their fur, like human hair, was highly radioactive. After all, the animals spent the whole day outside on the street and must have picked up vast amounts of radioactive particles.

Some dogs, finding themselves left behind by their masters, ran after the buses for a long way, but to no avail. Eventually they fell back and returned to the abandoned town, where they began to roam around in packs. First they devoured a large number of the radioactive cats and then turned wild and began to growl at humans. They even attacked humans and abandoned farm animals a number of times.

Then, for three days—27, 28, and 29 April (up to the day the government commission was itself evacuated from Pripyat to Chernobyl)—a hastily formed group of hunters with shotguns shot all the feral radioactive dogs. Their breeds included mongrels, Great

Danes, sheepdogs, terriers, spaniels, bulldogs, poodles, and lapdogs. On 29 April the hunt was completed, and the abandoned streets of Pripyat were strewn with the corpses of many different kinds of dog.

Archeologists once read an interesting inscription on some ancient Babylonian tablets: "When the dogs in a city band together in packs, that city will fall and be destroyed." The town of Pripyat was not destroyed, but it was abandoned and preserved as it had been by radiation for dozens of years to come—a radioactive ghost town.

The inhabitants of Semikhody, Kopachi, Shipelichi, and other nearby villages and farms were also evacuated. Anatoly Ivanovich Zayats, chief engineer at Yuzhatomenergomontazh, together with some assistants, including hunters carrying shotguns, went round to the farmhouses and explained to the people living there that they had to leave their homes.

It was painful to witness the anguish and tears of people who were faced with the prospect of leaving their ancestral lands for years to come, perhaps forever.

One old woman spoke up: "What's all that you're saying? Who's going to look after the house and the animals? What about the garden? Hey, sonny, who's going to do it all?"

Anatoly Ivanovich tried to explain: "You're going to have to, grandma. Everything around here is radioactive—the soil, the grass. You can't feed that grass to the cattle any more, and you can't drink the milk. None of it—it's all radioactive. The state will pay you compensation for everything, and everything will be fine."

But no one understood what they had heard, and they were not willing to understand.

"What do you mean? The sun is shining, the grass is green, everything is growing and the gardens are all in bloom. Look!"

"That's the whole trouble, grandma. Radiation can't be seen, and that's why it's dangerous. You can't take the animals with you. The cows, sheep and goats are radioactive, especially their coats."

Many local people, on hearing that they could not feed the grass to their livestock, drove the cows, sheep, and goats up ramps onto the roofs of their barns and left them there, to prevent them from eating the grass. They thought that it would be for only a day or two, and that things would return to normal.

It all had to explained over and over again. The livestock was shot, and the people were taken to a safe place.

TRYING TO SEAL OFF THE REACTOR

Let us now return to Pripyat and Air Force General Antoshkin.

On the morning of 27 April, in response to his call, the first two MI-6 helicopters arrived, flown by B. Nesterov and A. Serebryakov, two experienced pilots. The roar of the helicopter engines, as they landed on the square outside the Party Committee offices, awoke all the members of the government commission, who had not gone to bed until 4 A.M.

General Antoshkin was up on the roof of the hotel controlling the flight and landing of the helicopters; he did not sleep one minute that night. Nesterov and Serebryakov, carefully reconnoitering the whole of the site of the nuclear power station and its vicinity, worked out a flight pattern for the dumping of sand into the reactor.

The aerial approaches to the reactor were hazardous, especially on account of the ventilation stack of No. 4 unit, which was 490 feet (150 m) high. Nesterov and Serebryakov measured the radioactivity above the reactor at various altitudes; they did not descend, however, below 330 feet (100 m) where the radiation levels abruptly rose to 500 roentgens per hour. After each "bombing run," of course, those levels could be expected to increase significantly. In order to drop sand, the pilots would have to hover over the reactor for 3 or 4 minutes, which would be long enough for them to receive a dose of between 20 and 80 roentgens, depending on the level of background radiation. And how many sorties would there be? It was still hard to say. Things would become clearer as the day went on. They were truly at war—and it was a nuclear war.

Helicopters were already landing and taking off from the square in front of the Party Committee offices. The deafening roar was making it difficult for the members of the government commission to do their work, but they all made a special effort, by raising their voices and even shouting. Shcherbina was increasingly nervous, as he inquired why no bags of sand had yet been dumped into the reactor.

The rotor blades of arriving and departing helicopters blew up radioactive dust containing fission particles. In the air near the Party Committee offices and in nearby buildings, radioactivity rose sharply. Everyone was stifling.

Yet the destroyed reactor continued to belch more and more millions of curies of radioactivity into the sky.

Leaving Colonel Nesterov in his place as traffic controller on the hotel roof, General Antoshkin went up himself to inspect the reactor from the air. It took him a long time to pinpoint the exact location of the reactor. In fact, anyone unfamiliar with the design of the plant would have experienced similar difficulties. He realized that an operator or someone from the assembly branch would have to go along on the "bombing runs."

More helicopters were arriving all the time, filling the air with the uninterrupted roar of engines.

The reconnaissance flights had now been completed, and the approaches to the reactor identified. The next requirement was for bags, shovels, and sand, as well as people to fill the bags and load them onto the helicopters.

General Antoshkin put all these problems to Shcherbina. By now everybody in the Party Committee offices was coughing, had dry throats, and found talking difficult.

"What about all the people under your command?" Shcherbina inquired. "You're asking *me* all these questions?"

The general stood his ground. "Pilots don't load sand! Their job is to fly aircraft and steer properly. They must be absolutely sure they're right over the reactor, each time. Their hands cannot be shaking. It's out of the question for them to be shoveling sand and dragging bags!"

"Here, general," said Shcherbina, "take two deputy ministers, Shasharin and Meshkov, and let them do the loading, and get the bags, shovels, and sand. There's plenty of sand around here, the whole place is built on sand. Find some open ground not too far away, and get moving. Shasharin! Get as many builders and installers as you can. Where's Kizima?"

Testimony of GENNADY A. SHASHARIN, former deputy minister of energy and electrification of the USSR:

Air Force General Antoshkin did a great job; he's a very business-like and vigorous general. He kept pestering people all the time; he was relentless.

About 550 yards (500 m) from the party offices, near the Pripyat café down by the jetty, we found a large pile of good-quality sand which had been dredged out of the river for use in the construction of new apartment buildings. We found plenty of bags in a warehouse, and then the three of us—A. G. Meshkov (first deputy minister of medium engineering), General Antoshkin, and I started filling the bags. We were soon bathed in sweat. We were working in the clothes we had been wearing back at the office. Meshkov and I were wearing our Moscow suits and shoes, while the general was in his dress uniform. None of us had respirators or dosimeters.

I soon arranged for some others to help us, including N. K. Antonshchuk, the manager of Yuzhatomenergomontazh, A. I. Zayats, his chief engineer, and Yu. N. Vypirailo, the head of hydro-electric operations.

Antonshchuk came up to me with a list of applications for bonuses for the people who were going to be filling the sandbags, tying them up, and loading them aboard the helicopters. In those circumstances the whole idea seemed utterly ridiculous, but I approved it right away. It used to be standard practice to pay such bonuses to people performing assembly or construction work in contaminated areas at operating nuclear power stations. But here? Antonshchuk and those who had to do this work were clearly out of touch, failing to understand that the contaminated area in Pripyat was everywhere and that bonuses should really be paid to all the residents of the town. But I decided not even to attempt an explanation, and let everybody get on with the job.

We still did not have enough people, however. I asked Zayats to drive to the nearest collective farms and get help.

Testimony of ANATOLY IVANOVICH ZAYATS, chief engineer of Yuz-hatomenergomontazh:

On the morning of 27 April, we had to organize help for the helicopter crews loading sand into bags. We didn't have enough people. Antonshchuk and I drove around the Druzhba ("Friendship") collective farm, stopping at each house. People were out working on their individual plots, but many were working in the fields, as it was spring and sowing time. We started explaining that the land was no good anymore, that the mouth of the reactor had to be plugged up, and that help was needed.

It was really hot that morning. People were in a relaxed mood, just before the May Day celebrations. They just refused to believe us and went right on with their work. Then we went to find the director of the collective farm and the secretary of the party organization. Together we went out into the fields and repeated the same explanations to the people there over and over again. Eventually they began to take us seriously. We found between 100 and 150 volunteers, men and women, who then proceeded to work nonstop loading sand into the bags and the bags on board the helicopters. And all this without respirators or other protective gear. On 27 April, the helicopters, with our help, made 110 sorties, and on 30 April, 300 sorties.

Testimony of G. A. SHASHARIN:

Shcherbina was extremely impatient. With the helicopter engines roaring outside, he yelled at the top of his voice that we were lousy workers, that we were no good. He drove us like cattle—all of us, ministers, deputy ministers, generals, not to mention the others— telling us that we were very good at blowing up reactors, but useless when it came to filling sandbags.

Eventually the first batch of six bags was loaded aboard an MI-6. Antonshchuk, Deigraf, and Tokarenko, who had assembled the reactor, took it in turns to fly on "bombing runs," so as to guide the pilots precisely to their target.

Colonel Nesterov, a top air force pilot, flew the first sortie, approaching the damaged unit in a straight line at 87 miles per hour

(140 km/hr), and taking his bearings from the two 360-feet (110-m) ventilation shafts on the left.

He arrived over the crater of the nuclear reactor at 360 feet (110 m). The radiometer read 500 roentgens per hour. The helicopter hovered over the target—the opening, formed by the tilted, white-hot biological shield and the reactor vault. As they opened the door, they felt a surge of heat from below; it was carrying radioactive gas, ionized neutrons, and gamma rays. No one was wearing a respirator, nor was the helicopter protected underneath by lead plating. This was added later, after hundreds of tons of sand had already been dumped. The crew looked out of the open door and, staring down into the nuclear volcano and aiming with the naked eye, dropped the bag. And countless bags thereafter. There was no other way to do it.

The first twenty-seven crews—together with Antonshchuk, Dei-graf, and Tokarenko, their assistants—were soon unable to go on, and were sent to Kiev for treatment. At an altitude of 360 feet (110 m) radioactivity rose to 1,800 roentgens per hour after the bags had been dropped. The pilots were being taken ill in the air.

The fact is that sandbags dropped from such a height produced both a violent impact on the extremely hot core and—especially on the first day—sharply increased releases of fission fragments and radioactive ash from burnt graphite. All of that ended up in people's lungs. Uranium and plutonium salts had to be flushed from the blood of these heroes for several months; in fact, they had to have their blood replaced several times.

Within a few days the pilots, on their own initiative, began wearing respirators and slipping pieces of lead sheet beneath their seats, thereby reducing the crew's exposure to some extent.

Testimony of COLONEL V. FILATOV:

At 7 P.M. on 27 April, Major General N. T. Antoshkin reported to Shcherbina that 150 tons of sand had been dumped into the mouth of the reactor. He felt quite pleased with this performance, as the effort involved had been considerable. Shcherbina, on the other hand, was not impressed: "Not very good, general. For a reactor like that, 150 tons is peanuts. We've got to move much faster." He also lashed out at Shasharin and Meshkov, accusing them of inefficiency. He put Tsvirko, the head of Soyuzatomenergostroy, in charge of loading the sand.

Testimony of M. S. TSVIRKO:

On the evening of 27 April, when Shasharin and Antoshkin reported on the bag-dropping operation, Shcherbina shouted at them for ages, saying they had done a lousy job. Instead of Shasharin, he put me in charge of loading the sand. Instead of the place where the sand had been first collected, which was highly radioactive, we moved to a sand quarry 6 miles from Pripyat, so that the people doing the work would not be exposed unnecessarily. We got bags from a depot and from shops, first tipping out the flour, grain, and sugar. Later on, bags were brought from Kiev. On 28 April, we were given optical dosimeters, but they needed to be charged, and it seemed they hadn't been. My dosimeter constantly showed 1.5 roentgens; the needle never shifted at all, so I got another dosimeter, this time showing 2 roentgens, and not an iota more. At that point I spat with disgust and gave up looking. We must have picked up around 70 or 100 roentgens, at least.

Because of fatigue and loss of sleep, General Antoshkin could hardly stand up, and was discouraged by Shcherbina's reaction. But not for long. He once again swung into action. Between 7 and 9 P.M. he established good relations with all the department heads capable of supplying sand, bags, and manpower for the loading operation. They hit on the idea of using parachutes to increase efficiency. The canopy of each was filled with as many as fifteen bags, like a big shopping bag, then attached by its straps to the helicopter and carried off to the reactor.

On 28 April, 300 tons were dropped. On 29 April, 750 tons. On 30 April, 1,500 tons; and on 1 May, 1,900 tons.

At 7 P.M. on 1 May, Shcherbina announced that the volume to be dropped would be cut in half. There was reason to fear that the concrete structures supporting the reactor might not hold and that everything would collapse into the suppression pool, causing a thermal explosion and a massive release of radioactivity.

All in all, between 27 April and 2 May around 5,000 tons of friable materials had been dropped into the reactor.

Testimony of Yu. N. FILIMONTSEV, deputy director of the main scientific department of the Ministry of Energy:

I arrived in Pripyat in the evening of 27 April, very tired after my trip. I managed to get into the Party Committee offices where the government commission was working, and then went to the hotel to sleep. I had with me a pocket radiometer I had received as a gift at the Kursk nuclear power station before leaving to work in Moscow. It was a fine instrument, with a time-lapse feature. In the ten hours I was asleep, I received a dose of 1 roentgen. The radioactivity in the room must have been 100 milliroentgens per hour. At various points on the street, it ranged from 500 milliroentgens to 1 roentgen per hour.

We shall return to Filimontsev's testimony later.

5

REVIEWING THE DAMAGE: 28 APRIL–8 MAY 1986

At 8 a.m. on 28 April, I went to work and entered the office of Yevgeny Aleksandrovich Reshetnikov, head of the Central Directorate for Power Station Construction, a department of the Ministry of Energy, to report on my mission to the Krymskaya nuclear power station. This particular central directorate, known by the abbreviated title Glavstroy, dealt with the construction and assembly of thermal, hydraulic, and nuclear power stations. As its deputy head, I was in charge of the nuclear sector.

Although I am a technologist myself and worked for many years in nuclear power station operations, after my bout of radiation sickness I was not allowed to work with sources of ionizing radiation. Therefore instead of working in operations, I had switched to Soyuzatomenergostroy, an organization that handled construction and assembly, where I was in charge of the coordination of assembly and construction at nuclear power stations. This meant that I had been working at the interface between technology and construction. While working at Soyuzatomenergostroy, under Tsvirko, I was invited by Reshetnikov to transfer to the new directorate.

In other words, the decisive factor for me in my new job was the absence of contact with radiation, as my total dose had already amounted to 180 roentgens. Reshetnikov, an experienced and vigorous organizer of the construction industry, was passionately dedicated to his work, although his health was weakened by heart disease. He had worked for many years in the provinces building factories, mines, and thermal and nuclear power stations. However, he was not familiar with the technological side of nuclear power stations, and even less with nuclear physics.

As I entered his office, I started to brief him on my trip to the Krymskaya plant, but Reshetnikov interrupted me: "There's been an accident at the No. 4 unit of the Chernobyl nuclear power station."

"What happened? What caused it?" I inquired.

"It's hard to get through to them," he replied. "The phones at the plant have been disconnected. Only the high-frequency line is working, and not very well. The receiver is in the office of Deputy Minister Sadovsky. But the information coming through is rather vague. It seems there was an explosion of detonating gas in an emergency tank in the central hall, blowing the top off the central hall and the roof off the drum-separators, and destroying the main circulation pump room."

I asked whether the reactor was intact.

"No one knows. It seems it might be. I'm going along to see Sadovsky and find out what the news is. I wonder whether you could take a look at the plans and prepare a memorandum for a presentation to Dolgikh, the secretary of the Central Committee. Use layman's language. Sadovsky will be making the presentation, and, as you know, he is a hydraulic technician and doesn't understand the fine points of nuclear technology. I'll keep you informed. If you find out anything, let me know."

"It would be best to fly down there and take a look firsthand," I said.

"Not so fast. Far too many people have gone down there already. There's no one left in the Ministry of Energy to prepare material for the report. You can fly there after the minister comes back with the second team. I may go, too. All the best."

I went to my office, propped up the plans, and started examining them. The emergency standby water tank for cooling the servo-drives of the protection and control (SUZ) system is necessary in case the original cooling system fails. It is mounted in the outside wall of the central hall, between levels + 50 (115 feet) and + 70 (230 feet). It has a capacity of 29,000 gallons (110 m³) and is directly vented to the outside. If radiolytic hydrogen had accumulated there, it should have escaped through the vent. Given the magnitude of the destruction, it seemed unlikely to me that the tank had exploded. A more plausible scenario would be an explosion of detonating gas far below, in the drainage header, where water returns from the channels of the protection and control system and which is only half full. I pursued this thought further. If the explosion had occurred down below, the shock wave could have ejected all the absorber rods from the reactor, in which case there would have been a prompt neutron power surge, and the reactor would have exploded. Moreover, if Reshetnikov was to be believed, the destruction had been pretty enormous. So we had an explosion in the emergency water tank, which did not seem probable, and the roof had been blown off the central hall and the separator rooms. But that was not all. Apparently the main circulation pump rooms had also been destroyed. That could have been done only by an internal explosion, perhaps in the reinforced watertight compartment.

Such a scenario sent shivers down my spine. There was so little information, however. I tried to phone Chernobyl, but to no avail, as the line was dead. I phoned Soyuzatomenergo, where the director, Veretennikov, also knew nothing, or at least so he claimed. According to him, the reactor was intact and being cooled with water, but the radioactivity situation was bad. Apart from him, nobody was able to tell me anything that made any sense. Conjecture was the order of the day. At Soyuzatomenergostroy the official on duty told me that on the morning of 26 April they had spoken to Zemskov, the chief construction engineer, who had told them there had been a minor accident and that they did not want to be disturbed.

I clearly had little material for my report; so I prepared a memorandum based on the assumption that there had been an explosion in the emergency tank, and on the possibility of another explosion in the lower drainage header, resulting in a prompt neutron power surge and the explosion of the reactor. The explosion had doubtless been preceded by a release of steam through the relief valves into the suppression pool. That would account for the explosion in the reinforced watertight compartment and the destruction of the main circulation pumps.

As events later proved, I was not far from the truth. At least I was right about the explosion of the reactor.

At 11 A.M., a very nervous Reshetnikov announced that he had at last been able to contact Pripyat over the high-frequency line. Radiation over the reactor was running at 1,000 roentgens per second. I told him that figure was 100 times too high to be true. Perhaps it was 10 roentgens per second. In an operating nuclear reactor, radioactivity is 30,000 roentgens per hour, as in the center of a nuclear explosion.

"So the reactor has been destroyed?" I asked.

"I don't know," Reshetnikov replied mysteriously.

"It must have been," I said firmly, and as if talking to myself. "That means there was an explosion which wrecked all the communication lines."

I could imagine the full horror of such a disaster.

"They're dumping sand on it," Reshetnikov said, again mysteriously. "You're a nuclear physicist. What else would you advise them to dump in the reactor to choke it off?"

"Twenty years ago, when we had a prompt neutron power surge at a closed facility, we dropped bags of boric acid into the reactor vault from the central hall. That worked. In this case I imagine boron carbide, cadmium, or lithium should be used, as they're all highly absorbent."

"I'll pass that on to Shcherbina right away."

On 29 April, Reshetnikov told me that Deputy Minister Sadovsky, on the basis of our memorandum, had informed Dolgikh and Ligachov about what had happened in Chernobyl. Later news came in about the fire on the roof of the turbine hall and the partial collapse of the roof.

Over the next few days, it became perfectly clear at the ministry in Moscow that a nuclear disaster without precedent in the history of nuclear power had occurred at the Chernobyl nuclear power station. The Ministry of Energy immediately organized a massive transfer of specialized construction equipment and materials to Chernobyl via Vyshgorod. Cement mixers, cranes, cement pumps, cement-making facilities, trailers, trucks and bulldozers, as well as dry cement mix and other building materials, were taken wherever they could be found, and moved to the disaster zone.

I shared my misgivings with Reshetnikov: if the core collapsed beneath the concrete floor of the reactor vault and came into contact with the water in the suppression pool, a terrible thermal explosion, releasing massive amounts of radiation, would occur. To prevent that, the water in the suppression pool should be drained immediately.

"But how would we get inside the suppression pool?" asked Reshetnikov.

"If no access is available, you should fire hollow antitank shells into the concrete. They can penetrate armor plating on tanks, so concrete should be no problem."

This idea was passed on to Shcherbina.

On 29 April 1986, the government commission moved from Pripyat to Chernobyl.

Testimony of GENNADY A. SHASHARIN, former deputy minister of energy and electrification of the USSR:

On 26 April, I decided to shut down No. 1 and 2 reactor units. We started the process about 9 P.M. and had completed it by 2 A.M. on 27 April. I gave instructions for 20 auxiliary absorber rods, evenly distributed, to be inserted into the empty channels of each reactor core. If there were no empty channels, the fuel assemblies were to be removed and replaced by auxiliary absorber rods. In this way the operational reactivity reserve would be artificially increased.

On the night of 27 April, Sidorenko, Meshkov, Legasov, and I were sitting around and pondering the possible causes of the accident. We tended to blame the radiolytic hydrogen, but then, for some reason, it suddenly occurred to me that the explosion must have been in the reactor. We also thought a saboteur might have attached an explosive charge to the drives of the protection and control system and blasted them out of the reactor. This led us to the idea of a prompt neutron power surge. That same night, 27 April, Dolgikh reported on the situation. He wanted to know whether more explosions were possible. Measurements we had already taken around the reactor showed no more than 20 neutrons per square centimeter per second, later changing to 17 to 18 neutrons. This suggested that no reaction was taking place. Admittedly, we were measuring from a distance and through concrete. The actual neutron density is unknown. It was not measured from the air.

That night I calculated the minimum number of operational staff needed to service No. 1, 2, and 3 units and gave the lists to Bryukhanov for action.

On 29 April, this time at the meeting in Chernobyl, I said that all fourteen remaining units with RBMK reactors should be shut down.

Shcherbina listened to me in silence, and then outside the room after the meeting, he said to me, "Listen, Gennady, let's not say any more about this. Do you have any idea what it would mean to leave the country without 14 million kilowatts of installed capacity?"

We had people on duty around the clock in the Ministry of Energy and in my construction department, checking on the flow of cargo to Chernobyl and making sure that priority needs were being met. It appeared that there were no remote-controlled devices for collecting radioactive objects such as the fuel and graphite fragments which lay strewn all around the damaged reactor unit as a result of the explosion.

As there were no such devices in the Soviet Union, we agreed to pay a West German firm 1 million gold roubles for three manipulators to pick up fuel and graphite from the plant site. A group of our engineers, headed by N. N. Konstantinov, chief engineer of Soyuzatomenergostroy, flew immediately to West Germany to collect the devices and learn how to use them.

Unfortunately we were unable to do anything with them. They had been designed for use on flat surfaces, whereas the Chernobyl plant was strewn with debris. We then placed them on the roof to pick up fuel and graphite on the roof of the de-aerator, but the robots got tangled up in the hoses left there by the firefighters. Eventually the graphite and fuel had to be picked up by hand. But more about that later.

I spent 1, 2, and 3 May on duty at our construction department monitoring the flow of cargo to Chernobyl. At the same time practically no communication with Chernobyl was possible.

THE GOVERNMENT COMMISSION: 4–7 MAY 1986

Testimony of G. A. SHASHARIN:

On 4 May, we found the gate valve which had to be opened in order to drain water from the lower part of the suppression pool. There was little water in it. We looked into the upper pool through the hole of the reserve passage and found it empty. I got two diving suits and gave them to the soldiers, so that they could go and open the valve. They also used mobile pumps and flexible piping. The new chairman of the government commission, I. S. Silayev, offered a special inducement by promising a car, a dacha, an apartment, and benefits for the rest of their lives to the family of anyone killed in this operation who succeeded in opening the valve. Those participating were Ignatenko, Saakov, Bronnikov, Grishchenko, Captain Zborovsky, Lieutenant Zlobin, and corporals Oleynik and Navava.

On Saturday, 4 May, Shcherbina, Mayorets, Marin, Semyonov, Tsvirko, Drach, and the other members of the government commission arrived by plane from Chernobyl. At Vnukovo airport they were met by a special bus which took them to No. 6 clinic, except for Tsvirko, who arranged to be met by an official car and left on his own.

Testimony of M. S. TSVIRKO, director of the All-Union Nuclear Power Construction and Assembly Enterprise, Soyuzatomenergostroy:

When we arrived in Moscow, my blood pressure was very high; both my eyes were bloodshot. While everyone was assembling to take the bus to No. 6 clinic, I sent for my official car and went to my usual No. 4 medical unit (the Kremlin hospital). The doctor asked me why my eyes were red. I told him that high blood pressure was the probable cause. He took my blood pressure, which was 220 over 110. Later on I discovered that radiation drives up the blood pressure. I told him I been exposed to radiation in Chernobyl and asked him to check my condition. He replied that I was in the wrong place for such an examination, and that I should go to No. 6 clinic. I asked him to examine me anyway; and after a blood and urine test, he discharged me. At home I washed myself thoroughly. I had done the same in Chernobyl and Kiev. I wanted to lie down and rest, but the phone rang. They had been expecting me at No. 6 clinic and wanted me over there immediately. I went most reluctantly, announcing as I entered, that I was from Chernobyl and Pripyat.

I was told to go to the waiting room. A dosimetrist passed a sensor over me and told me I appeared to be clean. I had washed carefully before going there, and I have no hair. In No. 6 clinic, I saw Deputy Minister Semyonov, who had already been shaved completely like a typhoid patient. He was complaining that after he had lain on the hospital bed his head was more contaminated that it had been before. It seems they had been put in beds used previously by the firefighters and operators who had been brought in on 26 April with severe radiation sickness. It turns out that the linen had not been changed, so patients were contaminating each other through the bedsheets. I demanded to be discharged and soon went home, where I was able to rest.

The other new arrivals at No. 6 clinic were examined with the sensor, undressed and washed, and had their heads shaved. Everything was very radioactive. Shcherbina was the only one who refused to have his head shaved. Once he had been washed, he dressed in clean clothes and, with his radioactive hair, went home. Shcherbina, Mayorets, and Marin were treated separately at a medical center adjacent to No. 6 clinic.

Apart from Shcherbina, Tsvirko, and Mayorets, all members of the government commission were kept at No. 6 clinic for treatment between one week and one month. A new government commission, headed by I. S. Silayev, deputy chairman of the Council of Ministers, then flew to Chernobyl to replace Shcherbina and the previous members.

On 5 May, Chernobyl was evacuated. A group of hunters shot all the dogs in the town. There were more dramatic farewells between masters and pets. All residents and livestock were evacuated from within an 18-mile zone. The government commission moved its headquarters farther away, to Ivankov. The radioactivity in the air increased sharply.

Marshal C. Kh. Oganov trained at reactor unit No. 5 with his assistants in the use of hollow-charge armor-piercing shells. Officers and installers also took part, in preparation for the actual firing against the wall of the damaged reactor building on 6 May. Through the resulting hole, a pipeline would be inserted so as to feed liquid nitrogen beneath the foundation slab to reduce the temperature.

On 6 May, Shcherbina held a press conference at which he understated the level of background radiation in the vicinity of the damaged reactor building and in Pripyat. Why did he do this?

The chairman of the State Committee on the Use of Nuclear Power, A. M. Petrosyants, spoke in justification of the Chernobyl disaster, by saying, "Science requires victims." He thought this was a very intelligent remark, but it really sounded blasphemous and stupid. People were dying.

On the same day, Marshal Oganov detonated hollow-charge armor-piercing shells at the damaged reactor building. The charge was attached to the wall of the reactor auxiliary systems unit, and the fuse lit. They blew holes in the walls of three rooms, only to find their path blocked by pipework and machinery. In order to pass pipes through, the hole would have to be widened. At that point they thought it more prudent to try a different approach.

Kizima proposed another solution: instead of using explosives, they would cut their way through with a welder's electric arc from a transport corridor. Room 009 was the other side of the wall. Preparations began.

In order to reduce the burning of graphite and uranium hexafluoride and keep oxygen out of the core, nitrogen was fed under the reactor.

According to an assembler who had just arrived, on 1 and 2 May the radioactivity of the air in Kiev was around 2,000 doses—though these figures need, of course, to be verified.

On 7 May, a headquarters was set up in the Ministry of Energy in Moscow to provide immediate and long-term assistance to Chernobyl, with staff handling communications on the high-frequency line until 10 p.m. in the office of S. I. Sadovsky, first deputy minister.

At a meeting in the office of Deputy Minister Semyonov, I proposed that the damaged reactor building should be buried in earth thrown up by a directed explosion. This question was examined with the help of experts from the department of hydroelectric construction, and found to be impossible. The subsoil of Pripyat consisted mainly of sand, which was ill suited to directed explosions. Hard bedrock, such as did not exist in the area, was needed for this purpose. This was a pity. I would have nuclear power stations built on hard bedrock, so that if the need arose they could be engulfed in earth and turned into a modern version of a Scythian burial mound. Isn't a single human life more valuable than the most advanced reactor unit?

The first radio-controlled bulldozers arrived in Pripyat: the Japanese Kamatsu and our DT-250. The handling of these devices is quite different. The Soviet model is started up manually and controlled remotely; if the motor fails while operating in a highly radioactive zone, a man has to be sent to start it again, whereas the Japanese Kamatsu is started and controlled remotely.

The dispatcher phoned from Vyshgorod, where the hardware intended for Chernobyl was piling up, to say that a colossal amount of machinery had already arrived. There were large numbers of drivers, who were proving difficult to control. Accommodation and catering were posing real problems. Everyone seemed to be drinking, saying it was for purposes of decontamination. In Kiev and Vyshgorod, the radioactivity in the air was 0.5 milliroentgen per hour; and on road surfaces and asphalt, 15 to 20 milliroentgens per hour.

The dispatcher suggested dividing the drivers into groups of ten, with the most sensible one in charge. Unruly drivers should be sent back home, and henceforth the selection of personnel should be based on the need to have an uninterrupted reserve of people to replace those whose condition did not permit them to continue—that is, those whose dose had reached 25 bers.

From time to time the radioactivity of the air in Kiev rose sharply, doubtless due to plutonium, the transuraniums, and so forth. When that happened the headquarters staff had to move farther away, to new premises, leaving behind their bed linen, furniture, and other objects. Then, in their new location, they would start all over again anew.

When Nikolai Ivanovich Ryzhkov, the chairman of the Council of Ministers, visited the disaster area, the main complaint he heard was about the poor medical treatment. The prime minister blasted the minister of health and his deputies to kingdom come.

Unfortunately it became evident that we in the Soviet Union did not have the necessary specialized hardware for eliminating and confining nuclear disasters such as the one at Chernobyl: for example, an "underground wall" machine capable of digging sufficiently deep trenches, or robotic technology with manipulators, and so on.

Deputy Minister Semyonov returned from the meeting with Marshal S. F. Akhromeyev, defense minister of the USSR. He said that it had been quite a substantial gathering, with some thirty high-ranking military officers, such as V. K. Pikalov, chief of chemical warfare, attending. The marshal, in stern tones, had made it quite clear to all those present that the army was not prepared to engage in decontamination work, as it had neither the technical nor the chemical resources required for the purpose.

The truth of the matter is that no one was prepared for the Chernobyl phenomenon. For thirty-five years, academicians had assured everyone that nuclear power stations were even safer than the simplest of samovars. As experience has shown, any assessment of the scientific and technological revolution in general, and of nuclear energy in particular, must be based on accurate theoretical assumptions. And it must also, of course, be based on the truth.

On 7 May, the radiation situation in the disaster area, received by the Ministry of Energy secretariat on the high-frequency line, was as follows:

- In and around the vicinity of the nuclear power station: graphite (close up)—2,000 roentgens per hour; fuel—up to 15,000 roentgens per hour. In general, the background radiation around the unit: 1,200 roentgens per hour (in the direction of the pile of rubble).
- Pripyat—approximately 0.5 to 1.0 roentgen per hour (air). Roads, asphalt—from 10 to 60 roentgens per hour.
- Roof of the solid and liquid waste storage facility—40 roentgens per hour.
- Chernobyl—15 milliroentgens per hour (air); ground—up to 20 roentgens per hour.
- Ivankov (37 miles [60 km] from Chernobyl)—5 milliroentgens per hour.

Kizima, the construction manager, called from Chernobyl to complain that they were short of cars and vans. Drivers with cars—makes like Moskvich, UAZ Volga, Rafik—who had come from various construction projects were leaving of their own accord, in their own radioactive cars, after they had absorbed their dose of radiation. It was impossible to wash down all the cars. Within the cars radioactivity varied from 3 to 5 roentgens per hour. He asked for dosimeters, either optical or with counters; but these were in very short supply. In fact, they were being stolen by departing drivers as souvenirs. Organizing the dosimetric monitoring of builders and installers was Kizima's greatest headache. The dosimetric staff were so demoralized that they were not even looking after their own safety.

I was on the phone on 7 May to the civil defense people, who agreed to provide 2,000 optimetric dosimeter sets with battery packs, already charged, from their Kiev base. I passed on the information to Kizima and asked him to send a car.

Many ordinary citizens had been phoning and coming to the Ministry of Energy, asking to be sent to Chernobyl to help with overcoming the effects of the disaster. Most of them, of course, had no idea about the kind of work that would await them. But nobody seemed upset by the thought of radiation. They were talking as if 25 roentgens was perfectly acceptable; while others were quite frank about their desire to make money. They were under the impression that people working in the disaster zone were paid five times the normal wage.

Most offers of assistance, however, were unselfish. One demobilized soldier from Afghanistan said, "So what, if it's dangerous? Afghanistan wasn't exactly a picnic either. I want to do something for my country."

A draft government decree about Chernobyl was prepared, concerning measures for the elimination of the consequences of the accident (equipment, vehicles, chemicals for decontamination, bonuses for builders and installers). Minister Mayorets was to address a meeting of the Politburo that same day.

At 8 P.M. on 7 May, they decided to pour liquid concrete over the pile of rubble from the blast, in order to cement the fragments of fuel and graphite in place, thereby lowering the background radiation. Sixty welders were required to assemble the pipeline delivering the concrete solution. When Deputy Minister Semyonov ordered the director of Soyuzenergomontazh, P. P. Triandafilidi, to assign people to this task, the latter yelled angrily in reply, "We're going to fry our welders with radiation! Who's going to assemble the pipelines at the nuclear power stations now being built?"

This prompted a new order from Semyonov to Triandafilidi, this time to draw up a list of welders and installers and forward it to the Defense Ministry for mobilization.

With heavy rain forecast in the region of the Chernobyl plant, Silayev, the chairman of the government commission, issued instructions for the storm drains of the town of Pripyat to be diverted without delay into the cooling pond reservoir (previously they had emptied into the Pripyat River). He also ordered the entire headquarters of the government commission to go to the damaged reactor building to organize urgent measures to cover up the radioactive pieces of graphite and fuel expelled from the reactor by the explosion. I shall return to this matter in due course.

Many months of hard and hazardous work still lay ahead, in the midst of fierce radiation fields. And the tens of thousands of people working in those fields understood nothing at all about radiation.

VISITING THE SCENE OF THE DISASTER: 8 MAY 1986

At 10 A.M. on 8 May, I received a message from Reshetnikov instructing me to take the 3 P.M. flight from Bykovo airport to Kiev, and from there on to Chernobyl. My assignment was terse: I was to find out what was happening, assess the situation, and report on it. As he signed my travel authorization, Deputy Minister Aleksandr Nikolayevich Semyonov, asked me to check on the radiation fields: "When we were down there, no one really knew what they were talking about, and now they are lying and covering up. Look into it for me, will you? And when you get back, please explain the dangers of radiation for me, in layman's terms. Look at me, here I am sitting with my head shaved. My pressure is going through the roof. Could that be nuclear, perhaps?"

We took off from Bykovo around 4 P.M. We had waited a long time for Semyonov, who arrived an hour late with his assistant, whom he had recruited from the Electro-technical Ministry, while serving there as minister before moving to his present post in the Ministry of Energy.

Apart from me there were three other deputy heads of departments in the Ministry of Energy: Igor Sergeyevich Popel, from supply; Yulo Ainovich Khiesalu, from energy-related equipment; and V. S. Mikhailov, deputy head of personnel in nuclear construction. Mikhailov was a lively, sociable but temperamental person, with a penetrating and shrewd gaze—always brimming over with ideas and initiatives, some of which did not make much sense. He never seemed able to sit still for very long.

Khiesalu was a calm, quiet man who rarely said much but, when he did, had a pronounced Estonian accent. A charming and good person.

Popel was a vigorous man, with a broad face and a pleasant disposition.

All three were going for the first time into a high radiation zone.

They were naturally apprehensive and sought to reassure themselves by sounding relaxed. All the way to Chernobyl they pestered me with questions about one and the same issue: what radiation is like, what it consists of, how it can be absorbed, how one can protect against it, and how many roentgens one can be exposed to safely.

We were traveling in a YaK-40 aircraft chartered specially by the Ministry of Energy and adapted for high-ranking officials. It was divided into two compartments: one forward, for the most senior persons; and one in the rear, which was used by everyone else. This hierarchy was more strictly observed in the pre-Chernobyl era, as the disaster sharply democratized the atmosphere on such charter flights.

To the left of the forward compartment, the minister and his assistant sat facing each other in armchairs on either side of a small table. On the right side there were four pairs of armchairs, occupied by the deputy heads of the central directorates, the heads of production departments and services of the various branches of the ministry.

Of all the passengers on that flight, I was the only one who had worked for a long time in nuclear power operations. As for the minister, although he had already spent his first nuclear week in Pripyat and Chernobyl, had been irradiated, and now sat with his head completely shaved, he did not really understand what had happened. Semyonov took a superficial view of events and was incapable, without the help of expert advisers, of taking any decision of any substance on the range of issues involved. He was a round person, well fed, and even on the fat side; now he sat silently, saying not a word to either of his assistants in the compartment. A faint smile floated on his lips.

I watched him out of the corner of my eye and got the impression that he had been overwhelmed by the nuclear disaster which had suddenly descended on him, and was utterly at a loss. A message was written all over his face: "How did I ever get into this totally mysterious energy business? Why did I take it upon myself to deal with the construction and operation of nuclear power stations, about which I know nothing? Why did I ever leave my own beloved electric motors and transformers. Why?"

Of course the minister may have been thinking about something else, but he was clearly bewildered by the nuclear mess into which he had been precipitated. Bewildered, but not afraid. He was incapable of fear, because he failed to understand the dangers posed by the nuclear disaster. In fact, he was not prepared to concede that there had been a disaster. Just an accident. A small breakdown.

Another passenger on that flight was Kafanov, deputy head of special hydraulic construction. A tall, somber man with a puffy face, outwardly he exuded Olympian confidence; but it was also his first encounter with radiation.

I was sitting in the first row of seats, next to the window. Below we could already see the broad expanses of the Dnieper River. Flooding had only just finished in the area, and it was just as well that it had, because if the accident had occurred one month earlier, all the radioactivity released would have found its way into the Pripyat and Dnieper rivers.

A nervous Mikhailov sat behind me. He was worried about the uncertain future and anxious to have everything clarified ahead of time; so he asked me, in hushed tones, apparently so as not to disturb the minister, "Tell me, how much can you absorb and leave no trace? So that nothing would happen to you?"

"Take it easy," I whispered back. "I'll explain everything once we're on the ground."

Popel was also worried. His rich and precise tones could be heard from behind: "I have high blood pressure. I heard somewhere that radiation drives it through the roof. Why do I need it?"

Kafanov and Khiesalu were silent. From time to time I glanced at the minister, whose set smile remained perfect throughout the flight. His empty gray eyes, which bore a hint of astonishment, stared ahead inscrutably into the narrow space in front of him.

As we came in for our landing in Zhuliana airport in Kiev at 6 P.M., we flew low over the city. For the rush hour the streets were unusually empty, with few pedestrians. Where was everybody? I had frequently flown into Kiev from that direction, and had spent time there while I worked at Chernobyl, but had never seen it so deserted. It made me sad.

Finally we landed. The minister immediately went his own way in a ZIL limousine. He was met by Sklyarov, the Ukrainian energy minister, whose face was deathly pale, and the local party secretary. We ordinary mortals, however, were met by G. P. Maslak, the head of the supply department in the energy ministry, a lean, bald, pleasant man who extended a warm welcome.

The whole of our team, headed by Maslak, got into a blue minibus. Mikhailov and Popel immediately bombarded him with questions. After all, he was from an entirely new sphere, the radioactive Ukraine. It scarcely seemed possible, but there it was.

Maslak said that, according to the radio, the radioactivity in the air was 0.34 milliroentgen per hour, and much more on the asphalt; there had been no official word about that level, but he had heard that it was 100 times greater at ground level. He had no idea what all that meant, as this was his first contact with nuclear matters. He also told us that in the week since the explosion about 1 million people had left Kiev. In the first few days, the scene at the railroad station had been unimaginable, with bigger crowds than during the evacuations of the Great Fatherland War.* Speculators were charging up to 200 roubles for train tickets, despite the additional trains specially put on for the occasion. Passengers were storming onto the trains, leaving on the roofs of compartments, and hanging on to the steps. This panic, however, lasted only two or three days. Now it was possible to leave Kiev normally. Apparently the trouble started when high-ranking officials started quietly removing their children from the city. The effects were immediately discernible in the schools, as classes started to thin out.

Things were hard in local industry, as it was difficult to arrange two shifts per day, let alone three. However, those who had stayed behind—the absolute majority—were in high spirits and prepared to meet the challenge.

Mikhailov, hook-nosed, with the same kind of long, graying goatee beard once worn by Kurchatov, asked impatiently, "Damn it! What does 0.34 milliroentgen really mean? Tell us all about it, Grigori Ustinovich!"

"Yes, tell us!" they all demanded in a chorus, including Maslak, who was from Kiev.

---

*The Second World War.—Trans.

I had no choice but to explain what I knew.

"The maximum permissible dose for nuclear power plant operators is 5 roentgens per year. For the rest of the population it is a tenth of that—0.5 roentgen per year or 500 milliroentgens. Divide that by 365 days and you find that an ordinary mortal is entitled to absorb 1.3 milliroentgens per 24 hours. Those are the standards set by the World Health Organization. Now, on May 8 in Kiev, if official figures are to be believed, the level is 0.34 milliroentgen per hour, or 8.16 milliroentgens per 24 hours, which is six times higher than the WHO norm. According to Maslak, at ground level the daily dose is 300 times the WHO norm."

Our minibus was still traveling along streets which, at 7 P.M., were half deserted.

Maslak pointed out that, in the first three days after the explosion, radioactivity in Kiev had apparently been as high as 100 milliroentgens per hour.

"That means," I said, "that the overall 24-hour dose was 2.4 roentgens, or approximately two thousand times the WHO norm for ordinary people."

Mikhailov suddenly exclaimed, "Come on, Maslak! You're in charge of supplies, where are our dosimeters?"

"You'll be getting dosimeters in Ivankov. Everything is ready for you there."

Mikhailov began urging the driver to stop outside a liquor store. "We need vodka to decontaminate ourselves. Once your balls are irradiated, that's it. Is life worth living after that?"

The driver smiled but did not stop. Over the preceding ten days he had become convinced that he was not yet dead and that life was still possible.

"Of course not!" Popel exclaimed. "This is terrible! My pressure's already gone up. I've got a headache, and pain in the back of my neck."

"Pour some pee over it, it'll help," Mikhailov recommended.

"No, but joking aside," Popel went on, "what do they need me for? I don't understand anything. As soon as we get there, I'm going up to Sadovsky and saying, 'do you need me, Stanislav Ivanovich?' And if he says no, I'm going right back where I came from. Don't leave until we've sorted all this out," he said to the driver, who nodded.

"I shall also ask Sadovsky," said Khiesalu.

"Sadovsky himself is an ignoramus. He's a hydraulic expert," said Mikhailov.

Popel reminded him that he was, above all else, first deputy minister.

Glancing out of the window at the passers-by, I saw that most of them looked sad and worried.

We passed Shevchenko Square, the intercity bus station from which I frequently returned in the 1970s from missions on the regular bus to Pripyat, and we moved beyond the city limits of Kiev.

I gazed out at the tall pine groves on either side, and realized that everything here, too, was contaminated, despite the apparent cleanliness. There were far fewer people around, and they seemed gloomier, lonelier. Very little traffic seemed to be coming in the other direction, from Chernobyl.

We passed Petrivitsy and Dymer, with cottages and villages off to the side of the road. Very few pedestrians were anywhere to be seen. Some children with rucksacks were on their way home from school. A familiar scene, with familiar people, but now somehow different.

Previously there had been swarms of people everywhere and plenty of lively activity. Now everything seemed to have slowed down and thinned out. I felt sad and, despite myself, guilty. All of us in the nuclear power industry, including me, were to blame for what had happened to these perfectly innocent people. Part of the guilt was also borne by those few colleagues of mine who also realized the dangers posed by nuclear power stations to the community and the environment. We, with our knowledge, had failed to persevere in our efforts to convey that knowledge to people. We had failed to penetrate the solid wall of official propaganda on behalf of the alleged safety of nuclear power stations. I found these sentiments sweeping over me, try as I might to resist them. Once again, I thought back to Chernobyl, Bryukhanov, and the past fifteen years of nuclear power plants in the Ukraine, and the causes of the explosion.

The account of the events of 26 and 27 April which I gave in earlier chapters was prepared later, after I had visited Chernobyl and Pripyat, and painstakingly questioned large numbers of people, including Bryukhanov, the section heads and shift foremen, and other participants in the tragic events at the plant. I also found that my many years of experience in nuclear power station operations, my radiation sickness, and my stay at No. 6 clinic in Moscow in the 1970s all helped me explore the very complicated situation and reconstruct the precise course of events. No one person knew the whole picture, as each of the participants or eyewitnesses was familiar only with his own small piece of the tragedy. It is my duty, however, to give as complete and accurate a picture as possible. Only the full truth about the biggest nuclear disaster ever to happen on this earth can help people analyze this tragedy in depth, learn from it, and rise to a new level of understanding and responsibility for the future. This applies not only to the narrow circle of experts, but to everyone, with no exceptions.

Meanwhile, however, we were driving toward Chernobyl, armed only with the scantiest information about the disaster, that is to say the information I had been able to glean in Moscow between 28 April and 8 May.

At 8:30 P.M., our minibus was traveling along the broad and completely empty Kiev-Chernobyl highway, which only ten days before had been alive with traffic and headlights. We now had only six miles to go to Ivankov. Having already exhausted their questions about radiation and its biological effects, my colleagues were now tired and quiet. Mikhailov or Popel would occasionally say something like, "Let's face it," and then lapse into silence again.

I asked Maslak whether there was any protective gear for us in Ivankov. He believed there would be, and had in any case phoned about it. That evening the minister would, like us, spend the night in Ivankov, only in a rented cottage. Shasharin would be in an apartment. All the dormitories and accommodation at the electricity concern's premises in Ivankov were filled to overflowing. Workers had been evacuated from Chernobyl a few days previously after a sharp rise in radioactivity.

I said we should try to reach the Chernobyl headquarters that same evening. That would take an hour, and allowing for time to change and have supper, it would take an hour and a half. I felt it was important for us to attend that evening's meeting of the government commission. Maslak seemed uncertain about our prospects.

Our minibus pulled into the yard of the Ivankov electricity offices at 9:30 P.M. We got out and stretched our legs. We had a quick bite to eat in a wooden shack in the yard, at a small canteen for electricity workers. Maslak went off to find our protective gear and check on sleeping arrangements.

We waited about thirty minutes. Three workers who had just returned from Chernobyl were engaged in a heated discussion a few yards away. They were wearing cotton overalls, one of them white and the other two dark blue, with dosimeters in their breast pockets. The one in white, a tall, bald man, had taken his cap off and was pointing toward the northwest, to the evening sky, partly covered by a dirtyish haze, and shouting, "It's hot today—two thousand doses of plutonium. It's really stifling." He frowned, coughed, and wiped his wrinkled face with his cap.

"I'm itching all over," said another. "It's like an allergy."

The third man said, "It's my legs, especially around the ankles," and, pulling up the legs of his overalls and bending down, started scratching his purple swollen legs.

We also turned in the same direction. The sky was now sullen and sinister, and we were looking at it with the same apprehension you might find among front-line soldiers in wartime.

"Here, in the yard there's 5 milliroentgens an hour," said the bald man in the white overalls.

Our throats were beginning to smart. Mikhailov became excited. "Hear that? Five milliroentgens. I'm sure I'm allergic to this stuff." Then he asked me, "What's the 24-hour dose for operational staff?"

"Seventeen milliroentgens."

"Hear that? Three hours, and you've got your 24-hour dose, right there! How much are we going to pick up there?"

"We'll be all right. Don't panic!"

Maslak came back with some bad news.

"There's no protective gear, no dosimeters, and nowhere to spend the night. Everything is full. People are literally sleeping on top of one another. There aren't enough beds to go round, so people are sleeping on the floor. We're going back to Kiev for the night. We can't go to Chernobyl like this, they'll send us back. In the first few days people were wearing whatever they liked. I phoned Kiev and told them to deliver a bag with protective gear and dosimeters to the Kievenergo hotel, where you will be staying. At 6 P.M. tomorrow, the minibus will pick you up and take you to Chernobyl."

We had no choice, so we got back into the minibus and went to Kiev, arriving there at 11:30 P.M. At the Kievenergo hotel we found a huge bag containing dark blue cotton overalls, boots, and black wool berets. I was unhappy about the wool berets, as wool soaks up radiation; cotton would have been better, but there were none. It was better than nothing.

While my colleagues were signing papers, I went outside. Here, just as much as in Ivankov, the air had a sharp taste to it. In some places it must have been 3 to 5 milliroentgens per hour. In the lobby a moment before, I had heard over the radio that the level was 0.34 milliroentgen per hour. They were obviously understating the danger, but why?

Next morning was bright and sunny, with a temperature of 77°F (25°C). Mikhailov, Medvedev, Popel, Khiesalu, Kafanov, Razumny, and Filonov climbed into the minibus. Traveling via Vyshgorod, we saw the same scene as the day before: a quiet subdued Kiev, with the few pedestrians who were about rushing to work, completely engrossed in their own concerns.

As we left Vyshgorod, there was a dosimetrist on duty at a highway patrol post. We saw the same dosimetrists, with radiometers on their chests and the long sticklike sensors stationed at the highway patrol units in Petritsy, Dymer, and Ivankov. They were stopping vehicles and passing their probes over the wheels of the few cars coming from the direction of Chernobyl. They let us through. Near the dosimetric post outside Ivankov, we were stopped to have our papers checked, including our permit to enter the zone. All was in order. On the shoulder there was a light blue Zhiguli with its doors and trunk lid wide open, revealing a number of packages and some carpets. The bewildered owners, a man and a woman, stood next to the car.

"Where are these things from?" asked the highway patrolman, as the dosimetrist passed his probe over the packages.

"From Chernobyl. But everything is clean," said the man.

"Not entirely," said the dosimetrist. "Five hundred millibers per hour."

"What's this all about?" the woman complained. "Those are our things, and don't you take them away!"

We drove on, had breakfast at the electricity workers' canteen in Ivankov where we had been the day before, and then headed straight for Chernobyl.

On either side of the road, as far as the eye could see, there were deserted green fields; nor was there any sign of life in the villages and small towns through which we passed. Either the inhabitants were still sleeping, or they had left. Hens were rummaging in the dust, and some fifteen sheep were wandering, without a shepherd, along the road toward Chernobyl. A boy carrying his rucksack on the way to school stared at us for a moment as we drove by, all dressed in blue. An old woman was tugging at a recalcitrant goat. There was hardly anyone about. Our eyes began to smart and our lungs itched.

"Terrible air today," said the driver and pulled on his "snout." That was the name we gave to the nylon antidust respirators, which look rather like the severed end of a pig's snout.

We overtook a column of cement mixers which were being rushed to Pripyat with dry cement mix. As we entered the 18-mile (30-km) zone, we encountered a military patrol and a dose checkpoint. Some of the men were wearing respirators, while others were not. They checked our papers and our permit to enter the zone, and let us go on.

An armored personnel carrier passed us in the opposite direction, its driver looking sternly ahead and wearing a respirator. Breathing was now becoming more difficult, and our eyelids were smarting. After the driver, all the others put on their respirators, but for me. I found it somehow demeaning and just could not bring myself to submit to the damned radiation. A Volga carrying a minister passed us and drove through a patch of dust on the road, sending a cloud carrying 30 roentgens per hour over our minibus. Then I put on my respirator. The minister's Volga disappeared around a bend in the road, and once again we were alone on the road. Now and then we passed a lumbering cement mixer with its load of dry cement mix, and then it was quiet once more. Not a soul was to be seen anywhere on the vast expanses of fields and in the farms and villages. The plants were still a fresh green, but from experience I knew that quite soon they would turn dark, wilt and turn browner than pine cones. After a burst of growth the green shoots would become sickly and, like sheep's wool, they too, as the "hair" of the soil, would accumulate radiation. In fact, the total volume collected by the plants would be two or three times greater than that on the roads.

I found myself repeatedly having to answer a barrage of questions from my colleagues, and explain what radiation is and how it is absorbed into the body. I really wanted to say that it can be absorbed with anything you like to mention, as it was everywhere about us, as well as inside us, in the air we breathed. I refrained, however. Instead, I gave a scientific explanation, which was not really reaching my audience. The fact that they had practically forgotten my earlier explanations in Kiev did not surprise me, as I was the only person in the minibus who had ever had anything to do with radioactivity.

Popel complained of pain in the back of his neck.

"My blood pressure is up," he concluded. "Who needs it? I fought in the war and went through a lot. As soon as we get there, I'm asking Sadovsky whether he really needs me. The truth of the matter is that I could be a thousand times more helpful in Moscow than in Chernobyl. And a hundred times faster."

Now and then Mikhailov, Razumny, and Kafanov glanced into their dosimeters, on which the needle on a dial showed the quantity of roentgens received. We had been given crude dosimeters, with a scale up to 50 roentgens, whereas what we needed then was a more sensitive instrument measuring up to 5 roentgens, for example.

"The needle on mine has gone below zero, off to the left," said Razumny. "What a piece of junk! Just typical!"

"That's because instead of absorbing radiation, you are now emitting it yourself," joked Filonov. "You've already given off more than you got."

"Mine is right on zero," said Mikhailov. "But my eyes are burning, and my legs are itching." And he scratched his ankles furiously.

"It's because you're scared, Valentin Sergeyevich," said Razumny. "That way you might get not only allergies but also diarrhea."

We then came up to a street-washing machine which was squirting out a foaming solution on the road. As some of it splashed against the underside of our minibus, I recognized the familiar sickening smell of desorbent solutions. Actually, washing down asphalt in this way is about as useful as applying a poultice to a corpse, because radioactivity is easily absorbed into tar; in order to make asphalt clean again, you have to dig it up and start all over again, or at least cover the contaminated layer with a clean one.

There was not a soul to be seen, and no birds either, except for a crow flying low in the distance. It would have been interesting to check it for radioactivity, as feathers tend to be quite absorbent. A few miles farther on, we came across one more living creature—a skewbald colt, running toward us from the direction of Chernobyl along the shoulder, stirring up radioactive dust. It looked utterly lost and bewildered, as it neighed pitifully, looking for its mother. All the local farm animals had already been shot, but this one had miraculously survived.

It was doing its best to escape, carrying with it, of course, a thick radioactive coat. Nonetheless it kept on running. Who knows, it might be lucky.

As we came closer to Chernobyl, we saw military camps with tent settlements, soldiers, masses of hardware, including armored personnel carriers, bulldozers, and heavy tracked vehicles with manipulator arms designed for the removal of obstacles. These looked like tanks, but without gun turrets. And then more tent settlements, and troops everywhere. These were chemical warfare units of the Soviet Army.

We past through what looked like a ghost village, without a living soul in sight. The unusual silence was beginning to depress me. After that, more fields to right and left, and radioactive crops stretching off into the distance. Some chickens were pecking and rummaging in the radioactive dust.

We eventually arrived at Chernobyl. The sky was blue and cloudless, but with a slight haze. The asphalt was wet with decontaminant solution. There were armored personnel carriers everywhere, on the streets and along the sides of the road. Traffic was flowing between the headquarters of the various ministries and government departments situated around town. We drove along the main street.

"Where to?" asked the driver. "The regional Party Committee offices, or the polytechnical school, to see Kizima? That's where the management of construction at the Chernobyl plant is now located."

"The regional Party Committee offices, please," I said.

The soldiers on patrol were wearing for the most part "snout" respirators, but some wore the "petal" variety. Soldiers were sitting, smoking, on some of the armored personnel carriers, with the hatches open. A number of them had removed their respirators altogether for the purpose, while others had poked a hole in the respirator through which they passed their cigarettes. There were also pedestrians, wearing respirators. These were people who for some reason had no car and urgently needed to go to the headquarters of the coal or the transport ministry.

We drove up to the square where the regional Party Committee offices were located, and which was crammed with vehicles, mainly cars of various makes, buses, minibuses, and armored personnel carriers reserved for the members of the government commission. On the square, outside the party offices and near the parked vehicles, numerous sentries wearing respirators were standing guard.

After a while, all those cars and other vehicles would have to be buried, because a month or two of operations in this area would suffice to fill the interior of each with as much as 5 roentgens per hour or even more.

Ye. I. Ignatenko, deputy head of Soyuzatomenergo, met us at the front door, with two men I had never met. Ignatenko was just standing there without his cap, with his jacket wide open and his respirator hanging around his neck, and smoking.

"Hello! What about the radiation safety rules?" I said.

"Hello! You've arrived! Go and see Sadovsky," he replied.

"Is the minister here?"

"He is. He only just arrived."

A dosimetrist standing next to the door, with a radiometer on his chest, was passing a sensor-baton over the ground, switching between the ranges. I asked him what the readings were.

"From the ground, 10 roentgens per hour. In the air, 1.5 milli-roentgen per hour.

"What about indoors?"

"Five milliroentgens per hour."

I went inside, followed by Popel and Khiesalu, both of whom were anxious to notify Sadovsky of their arrival.

Along the corridor on the first floor, each room was occupied by a different organization. The names were pinned to the doors: Atomic Energy Institute, Gidproyekt, coal ministry, transport construction, NIKIET (the main designer of the Chernobyl reactor), the Soviet Academy of Sciences, and many others. I went into the dispatcher's office, where Sadovsky was tormenting Popel and Khiesalu: "What are you doing here?"

"We don't know ourselves, Stanislav Ivanovich!" said Popel with a hopeful note in his voice.

"Now you go right back where you came from! Today! You have a car?"

"Yes, we do, Stanislav Ivanovich!"

Popel and Khiesalu rushed to the minibus, beaming. Their one ambition in life—to get as far as possible from radiation—had just been realized. I was sincerely happy for them.

I also notified the first deputy minister of my arrival, and mentioned the assignment given me by Semyonov and Reshetnikov. Sadovsky went to the polytechnical school, where Kizima had his construction offices, about a mile and a quarter from the regional Party Committee offices.

I glanced into the room with the title Atomic Energy Institute on the door. Two desks were wedged together, facing each other near the window. Yevgeny Pavlovich Velikhov was seated at the left desk, and at the other Minister Mayorets, wearing the same kind of dark blue cotton overalls as I had on and with a wool beret on his shaved head. We must have got our protective gear out of the same package. Sitting nearby were V. A. Sidorenko, deputy chairman of the Nuclear Safety Committee and correspondent-member of the Academy of Sciences, Academician Valery Alekseyevich Legasov, Deputy Minister Shasharin, and Ignatenko. I went in and sat down on the empty chair.

Mayorets was arguing with Academician Velikhov: "Yevgeny

Pavlovich! Someone has to take over the organization of this whole thing. There are dozens of ministries working here at present. The Ministry of Energy is not able to coordinate them all."

"But the Chernobyl nuclear power station is yours," Velikhov countered, "so you must organize everything." Velikhov, wearing a plaid shirt opened over his hairy midriff, looked pale and exhausted. He had already received a dose of around 50 roentgens. "And in any case, Anatoly Ivanovich, you've got to be aware of what happened. The Chernobyl explosion was worse than any other nuclear explosion. Worse than Hiroshima. That was only one bomb, whereas here the amount of radioactive substances released was ten times greater, plus a half a ton of plutonium. Today, Anatoly Ivanovich, you have to count people, and lives."

I had great respect for Velikhov. Here, I felt, was an academician who was concerned about the public's health.

Later on I found out that the phrase "counting lives" had acquired a new meaning at the time. At the evening and morning meetings of the government commission, whenever they discussed ways of performing a particular task—such as collecting the fuel and graphite ejected by the explosion, entering the highly radioactive zone, or opening or closing some gate valve or other—Silayev, the new chairman of the government commission, would say, "To do that we're going to have to count two or three lives. And for that, one life."

These simple, matter-of-fact remarks had a sinister ring.

Velikhov and Mayorets continued their argument over who should be in charge of the situation.

I left the office. I was looking forward to seeing Bryukhanov and having a word with him. The dangers I had warned him about fifteen years before in Pripyat, when I worked at the Chernobyl plant, had now materialized. I had a lot to say to him—or, rather, a lot of anger and resentment to pour out at him. All those things had now actually happened. Yet in those days he was so cocksure and determined to have his own way that he ignored the hazards and dismissed the possibility of a nuclear disaster. And it had begun to look as if he might be right. For ten years the Chernobyl nuclear power station had been the best in the Soviet energy system, generating kilowatts over and above the plan, the few minor accidents easily covered up; there were honor rolls, proud banners, medals, prestigious awards, more medals, glory. And then the explosion.

I was seething with rage. I believed that of all the people present, he alone was to blame. Or at least more to blame than anyone else.

For the past fifteen years, it was his policy, and his ideology, that had been put into practice. Fomin, his chief engineer, was merely a pawn implementing that policy and manipulated by it according to circumstances. However, was it only Bryukhanov's ideology? Bryukhanov himself was nothing more than a pawn manipulated by the ideologues of the period of stagnation, which was now gone forever.

But who was he? Down the dimly lit corridor I could see a small, frail bare-headed man with curly gray hair leaning against the wall. He was wearing white cotton overalls; there was a look of dejection and embarrassment on his lined, powder-white face. He looked at me with red, hunted eyes.

I had walked past him, and then was suddenly struck by a suspicion. "Bryukhanov?" I turned around. "Viktor Petrovich?"

"Yes, that's me," said the man by the wall, in his familiar dull voice, and then he looked away.

At first I felt sorry for him, my anger and resentment completely vanished. The man I now saw before me was pathetic and crushed. He once again looked up at me.

We stared at each other for some time.

"Yes," he said at last, and looked away once more.

The strange thing is that at that moment I felt ashamed of being proven right. It would have been better if I had been wrong. In that case, however, everything would have continued as it always had. Being in the right had its price.

"You're not looking well," I said rather stupidly. It certainly was a stupid thing to say, because hundreds and thousands of people were at that very moment being irradiated because of the efforts of that one man. Even so, I could not speak to him in any other way. "How many roentgens did you get?"

"Somewhere between 100 and 150," said the man by the wall, in a that ever-so-familiar dull, hoarse voice.

"Where is your family?"

"I don't know. Probably at Polyeskoye. I don't know."

"Why are you standing here?"

"Nobody needs me. I'm just hanging around, like a piece of shit. I'm no use to anybody here."

"Where is Fomin?"

"He went crazy. They sent him away for a rest."

"Where to?"

"Poltava."

"What do you make of the situation here?"

"No one's in charge. They're all at sixes and sevens."

"I was told that you asked Shcherbina for permission to evacuate Pripyat on the morning of 26 April. Is that true?"

"Yes. But I was told to wait for him to get there, and not to stir up panic. There was a lot we failed to understand at the time. We thought that the reactor was intact. For me it was ghastly, the most terrible night."

"For everybody," I said.

"We didn't realize at first."

"What are we doing standing here? Let's go into one of these offices."

We entered an empty room next to Velikhov's office and sat facing each other across the table, again eyeball to eyeball. There was nothing to talk about, as everything was so clear. I found myself thinking that Bryukhanov had been a delegate to the twenty-seventh party congress. I had seen him on television. The TV camera had scanned the hall a number of times, trying to find his face, which in those days was extremely dignified, the face of a man who had reached the pinnacle of recognition. His was an authoritative face.

"You reported to Kiev on 26 April that the radiation situation at the plant in Pripyat was within normal limits?"

"Right. That's what the instruments we had at the time showed. Apart from that, I was in a state of shock. In spite of myself, I kept replaying the accident over and over in my mind, and comparing my past successes with my utterly hopeless future. I didn't really recover my composure until Shcherbina came. I wanted to believe that something could still be salvaged."

I took out a notepad to write something down, but he stopped me.

"Everything here is contaminated. There are millions of radioactive particles on the table. You'll get them all over your pen and notepad."

Minister Mayorets glanced round the door, and Bryukhanov, very much his old self now, leaped to his feet, forgot about me, and disappeared through the door.

A stranger, also with a powder-white face, then came in (100-roentgen doses cause spasms of the surface capillaries in the skin, making the face look as if it has been powdered). He introduced himself. He was one of the supervisors at the plant. With a bitter smile he said, "If it hadn't been for the experiment with the run-down of the generator, everything would have been the way it was before."

"How much radiation did you pick up?"

"I guess around 100 roentgens. My thyroid was giving off 150 roentgens per hour in the first few days. Iodine-131. It's a pity they didn't give people the things they needed. A lot of them are going through hell right now. They could have used plastic bags." And suddenly he said, "I remember you. You were working with us as deputy chief engineer in No. 1 unit."

"Your face is familiar, too. Where are your operational people now?"

"On the first floor, in the conference room and the room next to it. In the room that used to belong to the first secretary of the regional committee."

I said goodbye and went to the first floor, thinking as I went that with radiation levels so high outside, it might be advisable to place lead screens over the windows.

Before going into the conference room, I walked slowly along the corridor on the first floor to see which offices were there and who was occupying them. For the most part it was ministers and academicians; but there was one door with no sign on it. I opened it, looked inside, and saw a gray-haired man sitting at a table in a long room, with the blinds half closed. It was Silayev, deputy chairman of the Council of Ministers and formerly minister of the Aeronautical Industry, who had replaced Shcherbina here on 4 May.

He looked at me without a word, an authoritative gleam in his eyes, clearly waiting for me to speak first.

"These windows should be covered with a lead sheet," I said, without revealing my identity.

He remained silent, but the expression on his face became increasingly unfriendly. I closed the door and went to the conference room.

(I note here that lead screening was not placed on the windows of the government commission headquarters during Silayev's tenure as chairman, but was done much later, on 2 June 1986, when L. A. Voronin, deputy chairman of the Council of Ministers, Silayev's successor, was already in office. This was done in response to a sudden escape of radioactive contamination, which burst through the bags of sand and boron carbide which had been dumped on it.)

Some of the plant operators were sitting with their records on the stage of the conference room at the chairman's table, keeping in touch by means of a battery of phones with the bunker and the control rooms of units 1, 2, and 3, where skeleton crews were on duty in shifts, keeping the reactors in a cooled-down state. All of those sitting at the chairman's table had a guilty look about them—not the former confident bearing characteristic of nuclear power station operators during their glory days. All of them had tired, powdery-pale complexions and eyes inflamed by loss of sleep and by radiation.

In various places around the room, people were sitting in small clusters. These were the representatives of specialized branches, discussing questions for the meeting of the government commission.

I walked past the chairman's table, which had been turned into an improvised control panel, toward the window. Nearby, in the first row of chairs, I recognized an old friend, the head of the chemical branch, Yu. F. Semyonov. He was talking about the decontamination of equipment with a man I did not know, who was wearing protective gear. I later discovered that he was a foreman.

I had hired Semyonov, an experienced man with sound judgment, who had come to Pripyat from Melekess in 1972. He had been very anxious to work at the Chernobyl nuclear power station. For many years he had worked on systems for the decontamination of radioactive water. He was very pleased with his job at Chernobyl and generally content with his lot in life.

"Hello there!" I said interrupting his conversation.

"Oh! Good to see you! You seem to have come at the worst possible time."

"That's for sure."

Over the years since I had last seen him, Semyonov, whose face was also powdery-pale, had turned quite gray. His pitch-black sideburns were now white.

"Didn't you apply for early retirement two years ago? You wanted to quit the plant floor and find some clean work?" I inquired.

"Yes, that's what I wanted, but things dragged out. I had thought of going to Melekess with the family, but then this came up and here I am."

At this point the operators called for Semyonov. Ignatenko came into the room and, seeing me, came over to talk.

"If your story 'The Expert Opinion' had been published before the explosion," he said with a smile (he had written the preface to this story), "it would by now have become a collector's item. Your judgment was really sound—the reactor unit was blown up by an explosion of detonating gas."

"That's why they held it back," I said, "so as to prevent the author from becoming a prophet. In fact those were the very words: 'To be printed after the publication of the conclusions of the government commission.' So it will eventually be published."

"Yes, they really did a job here," said Ignatenko thoughtfully, looking out the window. "We're going to take a long time to get over it."

Near the window was a large bag full of soccer ball bladders, white with talcum powder.

"Why so many bladders?" I asked.

One of the operators sitting at the chairman's table replied with an embarrassed smile that they used them to take air samples. I inquired where.

"Well, everywhere. In Pripyat, in Chernobyl, and in the 18-mile zone."

He laughed when I asked whether they were meant to replace Turkin sampling chambers (a kind of plastic bellows, with a valve, which when extended draw in samples of air or gas for testing). It appears that the proper chambers were in very short supply, whereas these were plentiful. The operator explained, in response to my query, that they could be inflated with a pump or, if necessary, by mouth, as there was also, in the circumstances, a tremendous shortage of bicycle pumps.

"If you inflate them by mouth, you get an inaccurate measurement," I said. "When you breathe in, half of the radioactive substances remain in the lungs. They act like a filter. Each time you breathe in and out, the radioactivity builds up."

"What are we to do?" said the operator, with a laugh. "We already inhaled so much in the first few days that we hardly notice something as slight as this."

I went with Ignatenko into the next room, formerly the office of the first secretary of the regional party committee, which was practically filled by an enormous U-shaped table. Men in cotton overalls were seated at it. I knew some of them. The powdery-pale Bryukhanov was sitting, with a very detached air, at the head of the table. I found myself remembering this pose from the days of his greatest success—a kind of studied indifference, or aloofness, as if he had nothing to do with the business at hand.

"What a marshmallow!" I thought, recalling Kizima's trenchant verdict on the man from many years back: "You will never get a hard and fast decision out of him."

A number of photographs of the destroyed reactor taken from a helicopter were laid out on the table, together with the overall plan of the site and other papers. Ignatenko and I examined the best of the pictures. Bryukhanov pointed to a black triangle of irregular shape on the floor of the central hall, which was piled high with debris.

"That's the spent fuel storage pool," said Bryukhanov. "It's crammed with fuel bundles. The basin itself is now empty, as all the water evaporated. The residual heat will destroy the bundles."

"How many fuel bundles are there?" I asked.

"The basin is full, there must be around five hundred."

"How will you get them out of there?" asked Ignatenko. "We'll bury them together with the reactor."

A tall, well-built, middle-aged general in dress uniform came in and, addressing everyone in the room, said, "Can anyone here advise me, comrades? I am the commander of a group of army dosimetrists. We just cannot coordinate our movements with the builders or the operators, and we have no idea what to measure or where. We are not familiar with the design of the plant, or with the approaches to the highly radioactive zones. Someone is going to have to coordinate our operations."

Ignatenko said, "Work with Kaplun, the head of the plant dosimetry service. He knows everything. And bring these issues up at the meeting of the government commission. You new here?"

"Just got here."

"Then do as I said."

The general left. Time was slipping by, and I needed a car to get to Pripyat and the damaged reactor building. I asked Ignatenko for help.

"That could be tricky," he replied. "There's a tremendous demand for cars. I don't have one myself. Too many people are in charge here. Try Kizima."

I went down to the first floor to the dispatcher's office. Pavlov, the deputy director of technical construction in the Ministry of Energy, was on duty at the high-frequency phone.

"Do you have a car?" I inquired. "I'd like to pop over to Kizima's headquarters."

"I'm afraid not. Everybody made their way here as best they could. It's pretty chaotic around here. Sadovsky has gone somewhere in his Zhiguli."

"OK, I'll walk. So long."

I went out into the street. The sun was warm. Some nasty, sickly vapors were rising from the asphalt, which had been sprayed with decontaminant solutions. I walked up the street. Normally the trees would be filled with birds chirping to welcome the sun, but now it was quiet. And the foliage on the trees seemed peculiarly still and lifeless. Although it was not yet dead, it did not seem to be as alive and as vibrant as it would have been in clean air. The green of the leaves itself seemed artificial, as if they had been covered with wax to preserve them; in fact they gave the impression that they were standing still, trying to sense the ionized gas in the air around them. That air was emitting up to 20 milliroentgens per hour.

The trees were, however, still alive; somehow they managed to extract from that plasma whatever they needed for life. Cherry and apple blossom was all around. Ovaries were beginning to appear here and there. But blossom and ovaries alike were now accumulating radioactivity, from which there was no escape.

Near the fence of an abandoned yard, a girl of about twenty, wearing white cotton overalls, was breaking off branches of flowering cherry. She had already made quite a large bouquet. I asked her where she was from.

"From Eisk. I've come to help out in Chernobyl. What of it?"

"Oh nothing. There are plenty of suitors around here. Nice young soldiers, as many as you like."

"You can keep your suitors," she said with a laugh. "I came here to help." And she dipped her face into the bouquet.

"Those flowers are contaminated," I said.

"Come on!" she said dismissively, and started breaking off more branches.

I also broke off some branches thick with blossom and then went with them to Kizima. As I turned into a narrow street on the left, a cement mixer came rumbling by, stirring up clouds of radioactive dust. I pulled on my respirator and covered my head more fully with my cap. Dust carries between 10 and 30 roentgens per hour.

The construction department of the Chernobyl nuclear power station (the name seems outmoded now)—or put more simply, Kizima's headquarters—was located in the former polytechnic. The entrance and the whole building were swarming with people, some of whom strode about purposefully, while others clearly lacked a purpose; some stood, while others were seated. The constant arrival and departure of vehicles stirred up clouds of dust which rarely had time to settle. It was a quiet, windless day with a brutally hot sun. Most people left their respirators dangling around their necks, though some put them on when the dust rose. Broken-down cement trucks, cement mixers, and dump trucks were parked in a yard a hundred feet or so from the polytechnic. Most of them were in working condition, but so heavily contaminated that the workers using them would be needlessly exposed to high doses of radiation. The drivers in particular would receive up to 10 roentgens per hour while sitting at the wheel. The fact that equipment was being rendered unusable by radioactivity was a big problem.

Not far from the entrance to the polytechnic were two armored personnel carriers and a number of cars and minibuses, whose drivers were dozing behind the wheel or smoking nearby.

A dosimetrist went by with a radiometer on his chest, measuring the radioactivity of the dust with a sensor on a long stick. There was a tall lime tree with masses of foliage not far from the entrance, but no birds sang. A large blue fly buzzed in the hot rays of the sun.

Not all signs of life had vanished, as there were still flies. And not only bluebottles, but ordinary house flies swarmed all over the interior of the building. From the smell in the air I could tell that the toilets were not working properly. In the lobby a dosimetrist was taking readings from the khaki overalls of a short, very nervous man with the dark brown skin characteristic of a nuclear tan.

"Where have you been?" said the dosimetrist, applying the sensor to the man's thyroid.

"Near the pile of rubble. And also in the transfer corridor."

"Stay away from there in future. You're quite radioactive."

"How much did I get?" said the worker.

"Just stay away from there in future, like I said," said the dosimetrist and walked away.

I asked him to measure the radioactivity of the bunch of flowers.

"Twenty roentgens per hour. Get rid of it."

I went out into the street and tossed the flowers into the yard next to the radioactive machines. Then I went back inside and looked into a couple of rooms. Workers in blue and green robes were resting all over the floor after being exposed. In one room a young man, propped up on his elbow, said to another, "I feel as if I'd been flogged, I'm so tired. I feel very sleepy, but just can't get to sleep."

"Me, too," his companion replied. "That's my 25 roentgens showing up."

In the lobby of Kizima's office, a dispatcher was hard at work talking to Vyshgorod on the phone, evidently to A. D. Yakovenko, manager of Yuzhatomenergostroy.

"We need people for the shift!" he shouted into the receiver. "We need drivers. . . . The supervisor is right here, I'll put him on. . . . No need to? . . . Yes, they've all got their doses already."

Several rather nervous men came out of Kizima's office. I went in. Kizima was alone; he was opening a can of mango juice. The fabric of the "petal" respirator had left what looked like cobwebs on his cheeks.

"Good morning, Vasily Trofimovich!"

"Hi! Nice to see you people from Moscow!" he replied unenthusiastically. There was always an ironic edge to his greetings, and not just now. That's how I always remember him. At work he took a purely functional interest in anyone he was talking to. Small talk was not his specialty. He nodded toward the can of juice and pointed out that it contained large amounts of many different vitamins, which were helpful in overcoming the effects of radiation.

He gulped down his mango juice, his Adam's apple trembling slightly as he did. "You can see that I'm working as the site engineer."

The telephone rang and he picked it up.

"That's right, Kizima. . . . Yes, Anatoly Ivanovich. . . . The minister," he whispered to me, covering the mouthpiece with his hand. "Yes. Pencil and paper? I've got them right here. . . . A straight line at a 40-degree angle. . . . Then a vertical line. OK. . . . And then a horizontal line. . . . That makes a right-angled triangle. Is that it?" He listened for a minute longer and then hung up.

"As you can see, I'm the site engineer. Minister Mayorets is the chief site engineer, and Silayev, the deputy chairman of the Council of Ministers of the USSR, is the site manager. Complete confusion. They haven't a clue about construction. The minister just called to ask me to draw a triangle on a piece of paper." He handed me the piece of paper. "He made me draw the pile of rubble near the unit, telling me that I was to pour cement in there—as if I was a first-grader and understood nothing. And I walked around that pile of rubble on the morning of 26 April, as well as several times since then, with no respirator and no dosimeter. And, in fact, I've just come from there. And he calls me up and has me draw triangles! So I drew one! What next? Quite honestly I don't need any ministers or deputy chairmen. This is a construction site; it may be hazardous because of radiation, but it's nonetheless a construction site. I'm the site manager. Velikhov is the only scientific adviser I need, while the military can organize operations and keep order. And of course I need men. Our own have disappeared—I mean the labor for the construction job. And even management. More than three thousand people have quit without papers, without authorization to leave. The dosimetric service is not organized, and we're short of dosimeters and radiometers. Most of the optical ones we do have are out of order. I've been sending men off to work in a hazardous area with only one dosimeter for twenty-five people, and even that's broken. But even the broken ones work like magic. The men trust that little bit of metal and will not go to be irradiated without it. You have a dosimeter, let me have it. With that I can send another twenty-five men into action."

"I'll let you have it when I get back from Pripyat," I promised. "And what about the ones I arranged for you to get from civil defense, did you send a car to get them? Fifteen hundred outfits would help. Organize the service yourself. Don't just wait for things to happen. Get yourself an experienced dosimetrist from the office."

"Right. That's what we'll have to do."

A man entered. He was the supervisor in charge of delivering dry cement to the mixing unit from which wet cement would then be pumped to the pile of rubble.

"Vasily Trofimovich," he said, turning to Kizima, "We need drivers to replace those that can't work any longer. We're burning up our people. This shift has already had the full permissible dose, almost all of them have picked up 25 bers or more. They feel terrible."

"What about Yakovenko?" I asked. "Three days ago his dispatcher called Moscow to complain that the drivers who had been sent were unruly, spending their time boozing and hanging around. He had nowhere for them to sleep and nothing for them to eat."

"Well, he must be lying! I desperately need people!"

The supervisor left. Kizima said that he had a burning sensation in his chest and complained of a cough and headaches; it was like that all the time. I asked why he did not put lead screens over the windows and in the driver's compartment of his vehicles.

"Lead is harmful," he said emphatically. "It makes people worry and interferes with the job. I've already seen it happen. Lead we do not need."

The phone rang and Kizima picked it up.

"Right. . . . Yes . . . And what does Velikhov say? . . . He's thinking? Let him think. For the time being, stop delivering cement to the pile of rubble." He hung up. "Liquid cement geysers have started sprouting up. As the liquid cement lands on the fuel in the rubble, we're getting either a nuclear power surge or just a disruption of the heat exchange and a rise in the temperature of the fuel. The radiation situation is getting much worse."

Someone knocked on the door. A young major general came in, with three other officers, one colonel and two lieutenant colonels. He introduced himself as Major General Smirnov and said that he had been advised to seek Kizima's help.

"Have a seat. What can I do for you?"

"Our unit has come to guard the cooling pond. The water is highly radioactive."

"As radioactive as the water in the first loop of an operating reactor," said Kizima. "Water from the flooded underground compartments of the plant was pumped from fire trucks into the pond. The water in the pond measures one microcurie per liter."

"OK. So as to prevent saboteurs from blowing up the dam, sending all the contaminated water into the Pripyat and Dnieper rivers, I am placing sentries all along the perimeter of the dam, but we need some kind of shelter for them from the radiation."

"I suggest slabs of concrete," said Kizima. "We have reinforced concrete slabs over 6 feet long you can use. Stand them up on end, leaning at a slight angle to form a door and you have your sentry box. Shall I send down an order for some?"

"Yes!" said the general delightedly.

Kizima made the arrangements over the phone, and the officers left.

Then I phoned Moscow, asking them to send drivers without delay to replace the ones who had been exposed to radiation. I also spoke about it with Yakovenko, who promised that twenty-five replacements would be in Chernobyl the next morning.

"Vasily Trofimovich," I said to Kizima, "I have to go over to the damaged unit. Can you let me have a car for a couple of hours?"

"That's a hard one. Once they've got their dose, the drivers assigned here from nuclear construction projects leave in their own cars, without warning and without waiting for replacements. They also carry radioactive contamination away with them."

"An order for a new, additional consignment of cars was issued yesterday in Moscow. I'll check on it when I get back from Pripyat today. Can I have a car?"

"One of our foremen went for the day to Kiev. Take his Niva. It has four-wheel drive; it might be all right. Get a radiometer from the dosimetrists. They'll lend you one for a few hours." Kizima told me the number of the car, and the name of the driver—Volodya.

"I hope he's not easily scared."

"No, he's tough, he's just out of the army."

I left Kizima's office. When I told the dosimetrists who I was, they gave me a radiometer for a few hours, and I checked and recharged my own DKP-50 optical dosimeter.

Fortunately Volodya happened to have a special pass to get into Pripyat. In 10 minutes we were on the highway leading to the Chernobyl plant. I had driven along that road a hundred times in the 1970s, and later on, when I was already working in Moscow and was sent there on mission. The 11-mile asphalt strip from Chernobyl to Pripyat was unique in that there was a special hard shoulder of pink concrete, over 3 feet wide, on either side of the roadway. This was intended to protect the asphalt and prevent it from cracking. At the time we were pleased to be the only ones with such a road, as it meant we had to spend less money on highway maintenance. But now . . .

"What happens if the engine stalls right outside No. 4 unit?" Volodya asked suddenly, with evident irony. "We had that happen to us already—not outside the plant, but in Pripyat. The radioactivity isn't so bad there."

"You're just out of the army?" I asked.

"Six months or so," Volodya replied.

"Then no problem," I said. "If it stalls, you'll start it up. What branch were you in?"

"I was the driver for the commander of a regiment, in a four-wheel drive vehicle, a UAZ-469. Look, there's a dosimetric station. Those are chemical warfare troops," Volodya said.

A large green tanker truck, with various attachments—pumps, instruments and hoses, and so on—stood on the side of the road.

A Moskvich drove up from the direction of Pripyat and was stopped. The wheels and underside of the car, and the top of the trunk, were checked with a sensor. The passengers and driver were asked to get out; then the car was washed with desorbent solutions. The soldiers were wearing respirators, and tight cloth hoods covered their head and ears, with a large flap descending over their shoulders.

One of the soldiers, with a radiometer on his chest and a long stick-sensor, waved to us to stop. He checked our pass, which Volodya had stuck on the windscreen, and found it in order. The sensor, when passed over our Niva, showed background levels.

"You can go," he said. "But remember, your car will get contaminated where you're going. That Moskvich over there has 3 roentgens per hour, and washing won't get it off. Take pity on your car!"

"We have a radiometer," said I, pointing to the instrument. "And we'll be careful."

The soldier stared at me with his piercing blue eyes and seemed to be shaking his head uncertainly, as if doubting my word. He then slammed the door and waved us on.

Volodya accelerated, and the car shot forward with a whistling noise. I looked at the asphalt roadway bordered by pink concrete shoulders, and thought that we had rejoiced too soon, back in the days when the concrete had just been added, at not having to repair cracked asphalt any more. Now everything—including the asphalt and the concrete—was severely contaminated.

Thinking it might be interesting to see how fast the radioactivity increased as we approached Pripyat, I rolled down the window and held the sensor outside. On the right and straight ahead, beyond the radioactive vegetation flashing past, I could see the buildings of the Chernobyl nuclear power station, a bright white in the May sunlight, and the latticework structures of the high tension power masts of the 330- and 750-kilovolt switching stations.

I already knew that the explosion had ejected fragments of fuel onto the ground around the 750-kilovolt switching stations, where they continued to emit large amounts of radiation.

The pile of blackened wreckage visible outside No. 4 unit made a stark, painful contrast with all that elegant whiteness and latticework.

To begin with, the needle on the radiometer showed 100 milli-roentgens per hour, and then steadily crept to the right—200, 300, 500 milliroentgens per hour. Suddenly it shot off the scale. I switched through the ranges. What could that mean? Probably a nuclear gust from the damaged reactor building. A mile or so farther on, the needle dropped again, this time to 700 milliroentgens per hour.

The familiar old sign was now plainly visible in the distance: "Lenin Nuclear Power Station, Chernobyl" with a concrete torch. Beyond that, a concrete sign: "Pripyat, 1970."

We turned right, past the construction offices and the cement plant, toward the reactor unit straight ahead and then slightly to the left, in the direction indicated by the concrete arrow, the bridge over the railroad, with Yanov station on the left, and then into the town of Pripyat, where only recently 50,000 people used to live. But now . . .

"Volodya, let's go into Pripyat first."

He veered off to the left, accelerated, and soon we were crossing the bridge. Soon the snow-white town came into view, in the bright sunlight. As the needle on the radiometer had swung right on the bridge, I began to switch through the ranges.

"Let's get out of here—fast," I said. "The radioactive cloud passed this way and did some real damage. Faster!"

We shot over the hump of the bridge and raced into the streets of the dead town. A painful sight met our eyes immediately: the bodies of cats and dogs, everywhere, on the roads, in the yards, on the squares—white, brown, black, and spotted corpses of shot animals.

These sinister sights reminded us that this was an empty, abandoned town, to which normal times would never return. I wondered, nonetheless, why someone did not clean up. After all . . .

"Drive along Lenin Street," I said to Volodya. "It's easier to go by the house where I lived when I worked here."

It was number 9, I remembered.

Down the middle of Lenin Street there were young poplars, already quite tall, and on either side of the street, paths with benches and thick bushes. The imposing building of the Party Committee of Pripyat could be seen at the end of the street. To

239

the right of that was the ten-story Pripyat Hotel, and farther to the right, the jetty on the Pripyat River. Beyond that was a restaurant, and the road to the Lastochka Hotel, where visiting high officials used to stay.

The town looked really strange, as if it was quite early in the morning; but, in fact, it was bright daylight, with the sun high in the sky. Everything was in a deep sleep, from which it could not be awakened. There were household objects and laundry on the balconies and wilted flowers on the window sills. The sun's reflections in some of the windows had an unreal quality; one window had been left open, its curtain hanging out like a dead man's tongue.

"Stop, Volodya. Here, on the right. Slow down."

The needle of the radiometer crept in either direction, from 1 roentgen per hour to 700 milliroentgens per hour.

"Go slow," I asked. "There's my house. That's where I lived, on the second floor. Look how high that mountain ash has grown. Its blossom is all radioactive now. When I was here, it hadn't reached the second floor, but now it's up to the fourth."

The place was empty. The windows were all shuttered, but you could sense that there was no life behind the shutters. They were painfully still. There were some bicycles on the balcony, some boxes, an old refrigerator, skis with red poles. And no sign of life anywhere.

The body of a large black Great Dane with white spots lay across the narrow concrete path across the yard. I asked Volodya to stop nearby so that I could check the radiation level on its coat. He turned so that the left wheels went over a flower bed and stopped. Radiation had darkened the green leaves and made the flowers wilt. The ground and the concrete on the road measured 60 roentgens per hour.

"Look!" said Volodya, pointing to the three-story school building and the large windows of the gymnasium. "My son went there. I remember going to the school hall for special occasions, and all the kids and teachers looking so happy."

Two large but emaciated pigs were running toward us along a narrow path from the school, along the wall of a five-story building. They rushed toward the car and, whimpering, rubbed their snouts against the wheels and the radiator. They had a plaintive, hunted look in their bloodshot eyes, and their movements were shaky and ill coordinated. They were obviously extremely weak.

I held my sensor close to the side of one of the pigs—50 roentgens per hour; and then to the body of the Great Dane—110 roentgens per hour. The pig tried to catch the sensor in its teeth, but I pulled it away in time. The radioactive pigs started to devour the Great Dane. They easily tore off chunks of the partially decomposed flesh, shaking the body and dragging it along the concrete. A swarm of alarmed blue flies flew out of the open mouth and the decaying eyes.

"Just look at those flies! Aren't they something? Radiation has no effect on them! Let's go back, Volodya."

"Where to?"

"To the bridge and then on to the damaged reactor building."

"What if we stall?" said Volodya a second time, with a sly smile.

"If it stalls, you'll start it again," I said, with exactly the same tone. "Let's go."

Once we had turned onto Lenin Street, Volodya asked, "Shall we go in the wrong lane? Or what? We should be over there. Shall we drive round the square?"

"There's no need."

"It feels really funny. People get tickets for things like that."

"See any traffic anywhere?"

Volodya smiled grimly, and we drove quickly past the corpses of cats and dogs, on the wrong side of the road, toward the damaged reactor building. We went really fast over the railroad bridge. The radiometer reading was suddenly very high and then fell again.

We drove along the old road which runs past the power station construction offices, the residential construction plant, the Lisova Peniya restaurant, and the cement works.

On the right we could see the horrendous destruction that had occurred at the No. 4 reactor building. The smashed masonry and the pile of rubble were all severely charred. Streams of gas ionized by radiation were surging upward above the floor of what had once been the central hall, where the reactor was located. Amid the blackened wreckage, the drum-separators, which had been wrenched from their moorings and lifted sideways, looked curiously new and sinister as they reflected the bright rays of the sun.

We had another quarter of a mile to go to No. 4 unit.

"Switch on the four-wheel drive," I said. "We may well need extra traction around here.

"Look, Volodya!" Inside the fence, near the devastated reactor unit, and next to the pile of wreckage, some soldiers were walking, picking things up. "Turn right—Here—Drive to the waste storage building and stop right up against the fence."

"We're going to get fried," said Volodya, looking at me intently. His face was tense and red. Both of us were wearing respirators.

"Stop here. Uh-oh! I see officers, too. And a general."

"He's a lieutenant general," said Volodya, correcting me.

"Could be Pikalov. They're picking up graphite and fuel by hand. Look, they're going around with buckets and picking the stuff up. Then they're tipping it into containers—those iron boxes over there."

Graphite was also to be found lying about on the other side of the fence, right next to our car. I opened the door and placed the sensor practically on top of a graphite block. It read 2,000 roentgens per hour. I shut the door. There was a smell of ozone, burning, dust, and something else too. Perhaps scorched human flesh. After they had filled the buckets, the soldiers and officers walked over to the metal containers—in no particular hurry, it seemed to me—and unloaded their contents.

"You poor people," I thought to myself. "You're now gathering in a terrible harvest—the harvest of twenty years of stagnation. But what ever happened to the millions of roubles assigned by the state for the development of robots and remote-controlled manipulators? Where were they? Had the money been stolen? Or simply squandered?"

The faces of the soldiers and officers were the dark brown hue characteristic of a nuclear tan. Heavy rain had been forecast; and to prevent that rain from washing the radioactivity down into the soil, they sent men, instead of robots, because they did not have any robots. When Academician Aleksandrov found out about this later on, he was indignant: "They are showing no mercy for human life at Chernobyl. And I will be blamed for it." Yet he had not been indignant while proposing the installation of the potentially explosive RBMK reactor in the Ukraine.

Piles of sand could be seen some way off. Workers from the Ministry of Transport, who were digging beneath the reactor, had already built two tunnels. They were to be relieved later by coal miners.

"They're digging under the concrete foundations," said Volodya. "They tell me that under the reactor, a bottle of vodka sells for 150 roubles. For decontamination."

"Let's go, Volodya!" I ordered. "This road runs along the intake channel. You bear left."

We drove past the end wall of the turbine hall, where the reading was 200 roentgens per hour. Along the road past the transformers, I counted nineteen abandoned fire trucks.

Volodya turned onto the road. As we drove past the 750-kilovolt switching substations, the needle on the radiometer jumped to 400 roentgens per hour, most certainly because of fuel ejected by the explosion. Some 200 yards farther on, opposite the 330-kilovolt switching substations, the needle dropped to 40 roentgens per hour. And suddenly I gasped. Concrete blocks had been placed on the road, and there was no way through, as the railroad was on our left. It was a race against time, as the radiation level was rising every second.

"OK, Volodya, show us what you can do. Swing round onto the railroad tracks, stay on them for 50 yards or so, and then take the concrete road that leads to No. 1 administrative building. Let's go!"

The Niva performed beautifully, and so did Volodya.

Near the No. 1 administrative building, the level was 1 roentgen per hour. On the square in front of it, there were several armored personnel carriers, with a squad of soldiers in the middle. An officer was walking up and down reprimanding his men for violating radiation safety rules—by sitting on the ground, smoking, stripping to the waist to get a tan, drinking vodka, and so on. Neither the officer nor the men were wearing respirators, which were hanging about their necks.

It occurred to me that they were failing to follow the rules because they had not been trained properly. All of those young men would have children—but with even 1 roentgen per year, there is a 5-percent risk of a mutation.

"Stay here awhile, Volodya, I'll be right back. Be sure you don't go away, or I'll be stuck here."

Volodya gave me a sympathetic and reassuring smile.

Taking my radiometer, I dashed into the bunker, where there was no contamination, not even background levels. However, it was crowded and stuffy, rather like a bomb shelter in wartime. There were tables and beds on either side for people to relax. One group of men was playing dominoes; I could hear the pieces clicking against each other. Dosimetrists were on duty here, and operators were manning the phones, contacting the control rooms and the headquarters in Chernobyl and at the offices of the Party Committee. A map on the wall showed the radiation levels around the site; but I did not need it, as I had measured them myself.

I left the bunker and went up to the second floor of the administrative building. Here it was quiet and deserted. I took a corridor at level +10 (33 feet), the level of the de-aerator. From there on, I had to move quickly. My destination was the control room of No. 4 reactor unit. I had to see both the place where the ill-fated button was pressed, and how high up the needles showing the position of the rods had stopped. I had to measure the radioactivity in the control room and nearby, and understand the circumstances under which the operators had been working.

I practically ran along the enormously long corridor. It was over 600 yards to No. 4 control room. I had to hurry.

The radiometer read 1 roentgen per hour, but the needle was slowly moving to the right. As I passed control rooms 1 and 2, I could see the operators through the open doors. They were cooling the reactors—or, rather, keeping them in a state of cold shutdown. I reached No. 3 reactor, which had already been affected by the explosion, as levels were now 2 roentgens per hour. As I pressed onward, I noticed a metallic aftertaste in my mouth and a smell of ozone and burning, and could feel drafts. Shattered glass from the windows lay on the linoleum floor. Now it read 5 roentgens per hour. In the cavity near the Skala computer room, 7 roentgens per hour. I was now at the control panel of construction phase 2. I felt I was walking through the corridors and cabins of a wrecked ship. On the right were the doors to the stair-elevator well and, farther on, to the reserve panel. The door to control room 4 was now on the left. The men who were at that moment dying in No. 6 clinic in Moscow had been working right here. I went into the reserve control panel room, where the windows look out over the pile of rubble: 500 roentgens per hour. The windowpanes had been blown out by the blast, and crackled and squeaked as I walked on them. I went back toward control room 4. The reading at the entrance was 15 roentgens per hour; at the desk of the senior reactor engineer, Leonid Toptunov, who was then dying, it read 10 roentgens per hour. The needles indicating the position of the rods were stuck at between 6.5 and 8 feet. The farther I moved to the right, the higher the radioactivity; while at the far right end of the control room, it was between 50 and 70 roentgens per hour. I rushed out of the room and ran toward No. 1 reactor unit as fast as my legs could carry me.

The unthinkable had therefore happened: the peaceful atom in all its primeval beauty and horrendous power.

Volodya was still there. It was warm and sunny, around 85°F. The squad in the middle of the square had by now dispersed, and the officer had gone off somewhere. The soldiers were sitting around on armored personnel carriers, smoking; a couple of them had stripped to the waist and were tanning themselves. Young people believe they are immortal. They are immortal, as I could see for myself then and there.

I couldn't bear to look at them, so I shouted, "Hey fellows! You're needlessly exposing yourselves to radiation! Didn't he just explain all that to you?"

A fair-haired soldier smiled and stood up on the personnel carrier. "We haven't done anything. We're just getting a tan."

With that answer, I told Volodya to get going.

At about 8:30 in the evening, on 9 May, part of the graphite in the reactor burned through, forming a cavity beneath the materials that had been dumped into the reactor; and the whole enormous mass of 5,000 tons of sand, clay, and boron carbide collapsed, releasing a colossal amount of nuclear dust as it settled. There was a sharp rise in radioactivity at the plant, in Pripyat, and in the 18-mile zone. The increase was noticeable even in Ivankov and other places.

A helicopter was eventually sent up to measure the radioactivity, but only with great difficulty, as it was after dark.

The dust settled on Pripyat and the surrounding fields.

On 16 May, I flew to Moscow.

6

THE LESSONS
OF CHERNOBYL

THE PATIENTS AT NO. 6 CLINIC
IN MOSCOW

WHEN I PONDER the lessons of the Chernobyl tragedy, I think first of the hundreds of thousands of people whose destinies were to a greater or lesser degree affected by the nuclear disaster of 26 April 1986.

I think of the dozens of people killed, whose names we know, and also of the hundreds of unborn children, of lives cut short, of human beings whose names we shall never know because they died when their mothers' pregnancies were terminated by exposure to radiation in Pripyat on 26 and 27 April.

It is our duty to remember the outrageously high price that was paid for decades of criminal thoughtlessness and complacency on nuclear matters.

On 17 May 1986, in Mitino Cemetery, the civil defense service of the Ministry of Energy buried, with full honors, fourteen men who had died in No. 6 clinic in Moscow from injuries sustained on

26 April. They were operators and firefighters who had been at the damaged reactor unit. Doctors continued to fight to save the lives of others who had been seriously injured and of those whose condition was less dangerous.

Officials of the ministry took turns helping the medical staff at the clinic.

Early in the 1970s, I had lain there, on the ninth floor, under the supervision of Dr. I. S. Glazunov. At the time, the left annex of the clinic had not yet been built. My section was full of patients with severe radiation sickness, some of them extremely ill.

I remember Dima, a young man of about thirty, who had been irradiated while standing with his back, turned slightly to the right, only 18 inches from the source. The beam of radiation had hit him from below, doing most harm to his shins, the soles of his feet, and his perineum and buttocks, but tapering off toward the head. Having stood with his back to the source, he did not see the flash, but merely its reflection on the opposite wall and on the ceiling. When he realized what had happened, he rushed to turn something off, thus going one third of the way around the source. He spent 3 minutes in the danger zone. He reacted very calmly to what had happened, calculating his approximate dose. He was admitted to the clinic 1 hour after the accident.

On admission, his temperature was 102° and he was nauseated, chilled, and agitated, with glassy eyes. He gesticulated as he spoke, and tended to joke about what had happened to him, but was coherent and logical. His joking made some people feel uncomfortable. He was tactful, patient, and considerate.

Twenty-four hours after the accident, four samples of bone marrow were extracted from his sternum and his iliac bones (both front and left rear) for analysis. He was very patient during the puncture procedure. The average whole body dose was 400 rads. During the fourth and fifth days after the accident, he was in great pain from injuries to the mucous membranes of the mouth, esophagus, and stomach. There were sores in his mouth and on his tongue and cheeks; the mucous membranes were coming off in layers, and he lost both sleep and appetite. His temperature was now between 100.5° and 102°; he was agitated, and his eyes sparkled like those of a drug addict. Starting on the sixth day, his right shin, on which the skin was now disintegrating, began to swell and feel as if it was bursting; it then became rigid and painful.

On the sixth day, on account of a profound agranulocytosis (drop in the number of granular forms of leukocytes, responsible for immunity), about 14 billion bone marrow cells were injected (about 750 milliliters of bone marrow with blood).

He was then moved to a room sterilized with ultraviolet light. A period of intestinal syndrome began: bowel movements between twenty-five and thirty times every twenty-four hours, with blood and mucus; tenesmus, rumbling, and movement of fluids in the region of the cecum. Owing to severe lesions in the mouth and esophagus, he did not eat for six days so as not to irritate the mucous membranes. Nutrient fluids were provided intravenously.

In the meantime, soft blisters had appeared on the perineum and buttocks, and the right shin was bluish-purple, swollen, shiny and smooth to the touch.

On the fourteenth day he began to lose his hair, in a curious manner: all the hair on the right side of his head and body fell out. Dima said that he felt like an escaped convict.

He was very patient, but his jokes were wearing a bit thin. It was a kind of gallows humor, though he did much to cheer up his fellow patients who had been irradiated with him.

They were quite exhausted, although their condition was much less serious than Dima's. He would write them funny notes in verse, read Aleksey Tolstoy's trilogy *The Road to Calvary*, and say that at long last he had a chance to lie down. Sometimes, however, he lost his composure and plunged into a depression. The depression, too, was not particularly burdensome for his companions. Loud conversation, music, and the sound of heels used to irritate him for days on end. Once, while in such a depression, he shouted at one female doctor, saying that the noise of her heels had given him diarrhea. His family was not allowed in to see him until the third week.

On the fortieth day his condition began to improve, and on the eighty-second day he was discharged. He has a severe limp and a deep permanent sore on his right shin. Some thought had been given to the possibility of amputating his right leg at the knee.

The second patient, Sergei, aged twenty-nine, was alone in an adjoining sterilized room. He had been working at a scientific research institute, manipulating radioactivity substances in the "hot chamber." Two pieces of fissionable material had been brought too close together, causing a nuclear flash.

Despite instantaneous vomiting, he calculated his approximate dose—10,000 rads. Half an hour later, he lost consciousness and was flown to a hospital in extremely serious condition. Repeated vomiting, a temperature of 104°, and swelling of the face, neck, and upper extremities. His arms were so swollen that his blood pressure could not be taken with the usual cuff, and the nurses had to enlarge it.

He patiently endured the biopsy and bone marrow puncture. He was fully conscious. Fifty-four days after the accident, his blood pressure suddenly fell to zero. Fifty-seven hours later, Sergei died of acute myocardial dystrophy.

After I had been discharged, my doctor, with whom I had become quite friendly, told me about Sergei's death: "Under the microscope it was quite impossible to see the heart tissue, as the nuclei of his cells were nothing but a mass of torn muscle fibers. He really died directly from the radiation itself, and not from secondary biological changes. It's impossible to save such patients, as the heart tissue has been destroyed."

His friend Nikolai, aged thirty-six, who had been standing next to him at the time of the accident, survived for fifty-eight days. He was in agony the whole time, suffering from severe burns which caused his skin to flake off in layers, as well as from pneumonia and agranulocytosis. He received infusions of bone marrow in the old way, from sixteen donors. All these procedures did cure the agranulocytosis and the pneumonia. But he also had a serious case of pancreatitis which made him scream from the pain in his pancreas. Drugs did not help. Nitrous oxide anesthesia was the only thing that calmed him down.

It was early spring—in April, I believe—just as it was in Chernobyl at the time of the accident. The sun was shining and the hospital was very quiet. I looked in on Nikolai, who was alone in his sterile room. Next to the bed was a small table with sterile surgical instruments, and on another table were Simbezon and Vyshnevsky ointments, furacyllin, tinctures, creams, and gauzes, all of which were for use in the treatment of skin loss.

He lay on a high, slightly angled bed, above which powerful lamps shone down from a ribbed metal frame to keep his naked body warm. The cream had turned his skin yellow. But who is this? Nikolai . . . Vladimir Pravik. It was terrible to see how things are repeated! Fifteen years later, the same room, the same angled bed with the ribbed frame, heating lamps and ultraviolet lights turning themselves on at regular intervals.

Vladimir Pravik lay on his sloping bed beneath the lamps on their metal frame, the whole surface of his body a mass of burns, some caused by heat, others by radiation; in fact it was hard to tell one from the other. His body was swollen, outside and inside—lips, mouth, tongue, esophagus.

Fifteen years earlier, Nikolai had screamed from the pain in his internal organs and skin, but there had been no means to stop the pain. Now they have learned to do so. There has been much so pain over the years. But nuclear pain is especially merciless and unbearable. It induces shock and loss of consciousness. Even then, injections of morphine and other drugs soothed the pain of nuclear radiation syndrome for a while. Pravik and his comrades received intravenous bone marrow transplants and, by that same method, were given liver extract from large numbers of embryos in an attempt to stimulate blood formation. Yet death continued to advance.

He had everything possible wrong with him: agranulocytosis, intestinal syndrome, loss of hair, and stomatitis with severe swellings and detachment of the oral mucous membranes.

Vladimir Pravik stoically endured the pain and torment. This Slavic hero would have survived, would have overcome death, had it not been for the underlying death of his skin.

Most people in such circumstances would not be able to spare a thought for the common joys and sadnesses of life or for the fate of their comrades. Not Pravik, however. As long as he could still talk, he tried hard to find out through his sisters and doctors how well his friends were faring in their own struggles with death, whether they had survived. He so much wanted them to keep up the fight, so that their courage could help him. And when he somehow managed to hear news of certain deaths—perhaps foreshadowing his own—the doctors said that they had occurred somewhere else, in some other hospital. Those were creative, healing lies.

Eventually the day came when it was clear that everything of which modern radiation medicine was capable had already been done. All the methods of standard or more risky therapy had been used against acute radiation syndrome, but to no avail. Even the latest "growth factors," stimulating the multiplication of blood cells, did not work, because living skin was still necessary. Pravik had no skin at all, having lost it to radiation, which also destroyed his salivary glands and left him with a mouth as dry as soil in the midst of a drought. That was why he still could not talk. He just looked and blinked with his eyelids, which no longer had any eyelashes; he looked through his expressive eyes, in which a refusal to die was clearly discernible. Thereafter his inner strength began to ebb and eventually faded altogether. He began to wither and dry up with the approach of death, as the skin and body tissues of radiation victims do in fact become mummified. In this nuclear age, even death is transformed and made somehow less human, as the dead are blackened, shrivelled mummies, as light as a child.

Testimony of Viktor Grigoryevich Smagin, shift foreman in No. 4 unit:

In Moscow, at the No. 6 clinic on Shchukinskaya Street, I was put first on the fourth floor and then on the sixth. Those with the worst injuries—the firefighters and the operators—were on the eighth. The firefighters were: Vashchuk, Ignatenko, Pravik, Kibenok, Titenok, and Tishchura; and the operators, Akimov, Toptunov, Perevozchenko, Brazhnik, Proskuryakov, Kudryavtsev, Perchuk, Vershinin, Kurgus, Novik.

They were in individual sterile rooms, in which ultraviolet lamps came on periodically. The lamps were aimed at the ceiling so that their rays would not burn. All of us had a frightful tan—in fact, a nuclear tan.

The intravenous fluid we received at the Pripyat medical center made many of us feel better, as it removed the effect of the poisoning induced by the radiation. Patients with doses below 400 rads re-

... the improvement in other cases was
... ontinuing from skin burns due to heat
... he skin and internal organs was exhaust-

...?8 and 29 April, Sasha Akimov came into
... from his nuclear tan, and extremely de-
... ating that he did not understand why it had
... d been going well, and until the AZ button
... e parameters had deviated from the norm.
... than the pain," he told me on 29 April, as
... last time. We never saw him in our room
... nd stayed in bed. His condition suddenly got
mu...

All the seri... ill people were in sterilized individual rooms, on
high slanting beds, with heat lamps on above them. They had nothing on, because all their skin was inflamed and swollen; it had to be
treated, and the patients had to be turned over. All those with serious
or moderately serious injuries were given bone marrow transplants
and "growth factors"—medications that accelerated the growth of
bone marrow cells. Even so, the seriously ill ones could not be saved.

Testimony of LYUBOV NIKOLAYEVNA AKIMOVA, the wife of the shift
foreman of No. 4 unit:

Sasha's parents and twin brother took it in turns at his bedside.
One of the brothers gave some bone marrow for a transplant, but
it didn't help. While he could still speak, he kept on telling his father
and mother that he had done everything right, and that he could not
understand what had happened. This thought tormented him right
up to his death. He also said that he had no complaints about the
people on his shift, who had all done their duty.

I was with my husband a day before his death. He was already
unable to say a word, but you saw the pain in his eyes. I know he
was thinking about that damned fateful night, playing the whole
scenario over and over again in his mind, and he was unable to
accept that he was to blame. He had received a dose of 1,500 roentgens, perhaps more, and was doomed. His skin turned darker and
darker, and on the day of his death he was as black as a Negro. His

*body was really charred. He died with his eyes open. He and all his
staff were tormented by the same thought: Why?*

Testimony of V. A. KAZAROV, deputy director of Soyuzatomenergo:

*On 4 May 1986, I visited Slava Brazhnik, thirty years of age. I
tried to question him about what had happened, as nobody in Moscow knew very much at all. Brazhnik lay naked on a slanting bed,
all puffed up and dark brown, with a swollen mouth. With a great
effort he managed to say that he was in terrible pain all over his
body.*

*He said that, to start with, the roof fell in, and part of a reinforced
concrete slab fell onto the floor of the turbine hall, smashing an oil
pipeline. The oil caught fire. While he was putting out the fire,
another chunk of concrete came crashing down and wrecked a
feedwater pump. They turned that pump off and disconnected the
loop. Black ash came flying through the hole in the roof.*

*He found it painful to talk, so I asked no more questions. He
repeatedly asked for a drink. I gave him some Borzhom mineral
water. He told me that every part of his body was hurting him, and
that the pain was terrible.*

He said he never realized that such terrible pain was possible.

Testimony of V. G. SMAGIN:

*I went to see Proskuryakov two days before he died. I found him
lying naked on a sloping bed, his mouth grotesquely swollen and all
the skin on his face gone. Heat lamps shone down on him. He kept
asking for something to drink. I had some mango juice with me, so
I asked if he wanted some. He said he did, very much. He was fed
up with mineral water. There was a bottle of Borzhom on his
bedside table. I gave him some juice out of a glass. I left the can of
juice on his bedside table and asked the nurse to give him some. He
had no relatives in Moscow, and for some reason nobody at all went
to see him.*

*Lenya Toptunov, the senior reactor control engineer, had his
father by his bedside. He gave some of his bone marrow to help his
son, but nothing came of it. He spent all day and all night by his*

son's bed and turned him over. Lenya was burned black. Only his back was light, probably because it had been less exposed to radiation. He had been everywhere with Sasha Akimov, like his shadow, and they got burned the same way and almost at the same time. Akimov died on 11 May, and Toptunov on 14 May. They were the first of the operators to die.

Many patients who had thought they were getting better suddenly died. For example, Anatoly Sitnikov, deputy chief engineer for operations for phase 1, died suddenly on the thirty-fifth day. He had two bone marrow transplants, but they were incompatible, and his body rejected them.

Those whose condition was improving used to get together every day in the smoking room at No. 6 clinic, and the same nagging question kept coming up: "What caused the explosion?"

A host of possible scenarios were considered. They thought the detonating gas might have collected in the coolant drainage header of the SUZ protection and control system. Or, the control rods may have been ejected by an explosion from the reactor, resulting in a prompt neutron power surge. They also thought about the effect of the end sections of the rods. If that effect had combined with steam formation, that would have led to an power surge and an explosion. Everybody was gradually coming to the same conclusion—that there had been a power surge, but they could not be absolutely sure.

Testimony of A. M. KHODAKOVSKY, deputy director general of Atomenergoremont (the industrial unit responsible for maintenance at nuclear power stations):

The Ministry of Energy had instructed me to take charge of arrangements for the burial of those killed by radiation at Chernobyl. As of 10 July 1986, twenty-eight people had been buried.

Many bodies were radioactive. At first neither I nor the staff of the morgue were aware of this. Then we happened to take a reading and found high levels of radiation. After that, we started wearing suits impregnated with lead salts.

Once the epidemiological center had found out that the bodies were radioactive, they insisted on having a concrete slab underneath each coffin, just as in a reactor, to prevent the radioactive fluids from the body from entering the groundwater.

That was impossible, and blasphemous, and we spent a long time arguing with them. We eventually agreed that the most highly radioactivity bodies would be placed in lead coffins, which would then be soldered shut. That was how we did it.

Sixty days after the explosion, in July 1986, nineteen people were still being treated at No. 6 clinic. One man, whose condition was generally quite fair, suddenly broke out in burn marks all over his body on the sixtieth day.

The same thing happened to me. I had dark brown patches, of varying shapes, all over my abdomen; they were burn marks, obviously from working with radioactive corpses.

Testimony of V. G. Smagin:

Nikolai Maksimovich Fomin, chief engineer at the Chernobyl nuclear power station, was also a patient at No. 6 clinic, where he spent about one month. Shortly before his arrest, and after he had been discharged, I had lunch with him in a café. He was pale and depressed, with no appetite at all. He asked me what I thought he should do and whether he should hang himself. I told him I did not think that was a good idea and that he should be brave and stay with it right to the end.

I was in the clinic at the same time as Dyatlov. Just before he was discharged, he said to me, "I'm going to be brought to trial, that's for sure. But if they let me talk and if they listen to what I have to say, I'll tell them that I did everything right."

Shortly before his arrest, I met Bryukhanov, who said, "I'm no use to anyone. I'm waiting to be arrested. I went to see the public prosecutor to inquire what I should do and where I should be."

"And what does the public prosecutor say?"

"He says I should wait, and I will be called."

Bryukhanov and Fomin were arrested in August 1986, and Dyatlov in December.

Bryukhanov was quite calm. He had with him in his cell manuals and textbooks from which he was learning English. He said that, like Frunze,* he was condemned to death.

Dyatlov was also calm and composed. Fomin lost control completely, he was hysterical. He tried to commit suicide by smashing his glasses and cutting his veins with the pieces of glass, but they reached him in time and saved his life.

The trial, which had been postponed because of Fomin's insanity, was scheduled for 24 March 1987.

I eventually found and met with Razim Ilgamovich Davletbayev, the deputy head of the turbine section of No. 4 unit at the Chernobyl plant. As we have seen, he was in the No. 4 control room at the time of the blast. During the accident he picked up a dose of 300 roentgens. He looked really sick, with a severely swollen face and unhealthy, bloodshot eyes. He also had radiation-induced hepatitis. But he was keeping himself in shape and was in good spirits. He sported a stylish chestnut-brown mustache. Despite his disabilities, he was still working—a brave man.

I asked him to describe events on the night of 26 April. He told me that he had been forbidden to talk about technological matters, and that he had to go through the KGB. I replied that I knew everything about technology, even more than he did. What I needed was details about the people.

But Razim Ilgamovich was reluctant to say anything, and seemed apprehensive about the KGB all the time he was talking: "By the time the firefighters entered the turbine hall, the operators had already done everything. During the emergency in the turbine hall, from 1:25 to 5 A.M. on 26 April, I rushed over to the control room several times to report to the shift foreman. Akimov was calmly giving out orders. When it started, everyone reacted calmly. Being ready for events like that is part of our job, after all—though not for anything quite that bad, of course."

Davletbayev was nervous and obviously trying hard to stay within the limits set by the KGB, so I did not interrupt him.

---

*Mikhail Frunze, Trotsky's successor in the Commissariat of War. In November 1925, he fell ill; although surgery seemed risky to some of his doctors, he obeyed the Politbureau's orders, had an operation, and died while it was in progress. Trotsky suggested that Stalin had condemned him to death.—Trans.

He then gave a description of Aleksandr Akimov, his supervisor on the shift: "Akimov is a decent and honest man. He's friendly and sociable. He's a member of the Pripyat Party Committee and a good comrade."

He declined to offer a description of Bryukhanov, claiming that he did not know him. He did have opinions about the press coverage of Chernobyl: "I followed the press closely. They made people like me, on the operational staff, look incompetent and ignorant—in fact almost like villains. That is why all the photographs were torn off the tombstones at Mitino Cemetery, where our boys are buried; it was the influence of the press. The only photo they spared was Toptunov's. He was so young and inexperienced. But we were painted as villains. But for ten years the Chernobyl plant produced electricity—and that's not an easy way to make a living, as you know yourself. You once worked there."

"When did you leave the unit?" I asked him.

"At 5 A.M. I began throwing up violently. Even so, we managed to do everything. We put out the fire in the turbine hall. We got the hydrogen out of the generator and replaced the oil from the turbine oil tank with water.

"We were not merely agents, carrying out orders from higher-up. We figured out plenty of things for ourselves. But to a great extent, the horse had already bolted. I mean the technological process when we started our shift. And it just couldn't be stopped. But we were not merely agents."

THE LESSONS OF CHERNOBYL

Davletbayev was actually right about many things. Nuclear operators are not merely agents. In the process of operating a nuclear power station, they have to take vast numbers of independent and crucial decisions, sometimes involving great risk, to save the reactor unit, to emerge unscathed from a crisis or from a tricky transitional regime. Unfortunately no instructions and regulations can encompass the enormous variety of conceivable combinations of regimes and mishaps that may occur. Therefore the experience and thorough professional sense of the operators are vital. Davletbayev was right to say that after the explosion the operators performed miracles of heroism and courage. They deserve our admiration.

Nonetheless, at that one fateful moment before the explosion, the professional sense and the experience of Akimov and Toptunov deserted them. Both of them proved to be mere agents, although they did try—not very vigorously, it is true—to stand up to Dyatlov's bullying. That was the moment when the operators' professional sense came into play, but was overcome by their fear of being reprimanded.

No professional sense was shown by the experienced and cautious Dyatlov; by Rogozhkin, the plant shift foreman; by the chief engineer, Fomin; or by the director, Bryukhanov.

Whereas courage and fearlessness became the main driving force for the nuclear operators after the explosion, Dyatlov and Bryukhanov failed to behave professionally or honorably even then. Their self-serving lies and their wishful thinking continued to mislead everyone for a long time and resulted in further loss of life.

What, then, in my opinion, is the main lesson to be learned from Chernobyl?

Above all else, it is that this horrible tragedy summons us forcefully to the Truth—to tell the truth, the whole truth, and nothing but the truth. That's the first thing. My second conclusion derives from the truth.

Regardless of any measures that may be taken, a shutdown due to a positive excess of reactivity—in other words, an explosion—continues to be a possibility because of inherent flaws in the design of the RBMK reactor. This is because the reactor continues, as in the past, to have positive temperature and void coefficients, as well as the positive reactivity built into the ends of the absorber rods. The combined weight of these factors is excessive. It is no easy matter for all three to converge at the same time, but it can happen. They did converge at Chernobyl, and the consequences are there to see.

Like all the tragedies of the past, Chernobyl showed how great is our people's courage and how strong its spirit. But Chernobyl calls on us to use our reason and our analytical powers, so that we will not forget what happened, and will look clearly at our misfortune and avoid glossing over it.

Of course, some correct decisions have been made with respect to nuclear power stations with RBMK reactors:

- The tip switches of the SUZ protection and control rods are to be modified so that, in the fully withdrawn position, the absorber rods will still be inserted in the core to a depth of 47.25 inches (1.2 m).
- This measure will increase the speed of effective protection and preclude a constant increase of reactivity in the lower part of the core as the rods are lowered from the withdrawn position.
- The number of absorber rods permanently inside the core will be increased to between 80 and 90, thereby lowering the core's reactivity void coefficient to a tolerable level. Eventually this temporary measure will be replaced by converting the RBMK reactor to fuel with an initial enrichment of 2.4 percent, with the installation of fixed additional absorbers inside the core, so that the positive release of reactivity in an emergency will not exceed one β. In Chernobyl at the time of the explosion, it was five β or more.
- And, lastly, it has been decided that nuclear power stations with the RBMK reactor shall be gradually closed down and replaced by thermal power stations using gaseous fuel. This is certainly the most obvious measure taken as a result of the lessons of the Chernobyl tragedy.

It is to be hoped that these decisions will be carried out, because man, as a being endowed with reason, must ensure that all his scientific and technical achievements, especially nuclear power, are used to make life flourish, not fade.

Accordingly, the main lesson of Chernobyl is to sharpen our sense of the fragility and vulnerability of human life. Chernobyl demonstrated both man's immense power and his impotence. And it served as a warning to man not to become intoxicated with his own power, not to take that power lightly, and not to seek in it ephemeral gains and pleasures and the glitter of prestige. Since man is both the cause and the effect, he must be more responsible and scrutinize himself as well as the things he has made. When we remember that man's works carry over into the future, with all its joys and hardships, we realize with horror that those shattered chromosome strands and those genes, either lost or distorted as a result of radiation, are already part of our future. We will be seeing them again and again in the years ahead. That is the most horrible lesson of Chernobyl.

And those who died early, almost immediately after the explosion, those who suffered the agonies of nuclear death: Their memories haunt us. We would dearly like to see them again. The number of those who lie buried may not be large, but the magnitude of the pain and suffering they endured would suffice for millions. They carry within them and symbolize thousands and millions of deaths, leaving behind them on earth a painful and stern warning.

We bow to the martyrs and heroes of Chernobyl.

Testimony of Yu. N. Filimontsev, deputy director of the scientific research department of the Ministry of Energy:

After Chernobyl we went to the Ignalinskaya nuclear power station. There, in the light of the Chernobyl accident, they checked the design and the physics of the reactor. The total of the positive reactivity coefficients there is even greater than at Chernobyl; at any rate, it is not lower. The reactivity void coefficient is 4 β. They are doing nothing about it. We asked them why they did not write to the authorities. Their answer was that it made no sense to do so, it was pointless.

Nonetheless, the conclusions of the commission on the redesigning of all RBMK reactors with a view to enhancing safety have been unswervingly implemented.

The results of a number of investigations have been submitted to the government, among them the records of the Ministry of Energy, the government commission, and the Ministry of Medium Machine Building. All the external organizations have voiced conclusions unfavorable to the Ministry of Energy. Their main thrust was that the operational staff was to blame, and that the reactor itself had nothing wrong with it. However, the conclusions of the Ministry of Energy were more balanced and more thorough, pointing to faults on the part of the operators and flaws in the design of the reactor.

Shcherbina assembled all the commissions and demanded that they prepare an agreed set of conclusions for the Politburo of the Central Committee of the Communist Party.

MITINO CEMETERY

On the first anniversary of the Chernobyl disaster, I went to Mitino Cemetery to pay tribute to the memory of the firefighters and nuclear operators who lost their lives. A twenty-minute ride on the 741 bus from the Planernaya subway station took me to this huge city of the dead, just past the village of Mitino, on the outskirts of Moscow.

The cemetery was new and clean. Graves stretched off into the distance. To the left of the entrance was a neat building clad in yellow ceramic tiles; this was the crematorium. It was constantly in operation, with a faint black smoke coming from its chimneys. The cemetery office was on the right.

This was a young cemetery. The trees planted on the graves had not yet grown very tall. As it was still spring, they stood there leafless and dark. Flocks of crows swooped down on various parts of the cemetery to peck at the eggs, sausages, candies, and other food that had been left on the graves.

I walked along the main path. Some fifty yards from the entrance, to the left of the path, were twenty-six white gravestones. Over each one there was a small marble stele with an inscription in gilt lettering, stating a name, first name and patronymic, and dates of birth and death.

The six firefighters' tombs were piled high with flowers, including vases and pots with live flowers, wreaths of artificial flowers with red sashes, and inscriptions from family and colleagues. Soviet firefighters remember their heroes.

There were rather fewer flowers on the graves of the operators, and no wreaths at all. The Ministry of Nuclear Energy and the Ministry of Energy did not remember the dead on the anniversary of Chernobyl. But they, too, were heroes, they did all they could. They were fearless and brave, and gave their lives.

There were others in the cemetery who had just happened to be at the site of the tragedy on that ill-fated night and who never understood the real meaning of what was happening.

The sun shone from a clear blue sky. It was warm. The crows made their croaking call as they flew to and from the graves; people went on their way to the graves of their loved ones, along the main thoroughfare of the cemetery, which stretched off into the distance.

Not far from the graves of the victims of Chernobyl, I heard the noise of rifle fire. I glanced over and saw a squad of soldiers firing salutes with Kalashnikovs. A man passing by told me that they were burying a soldier killed in Afghanistan.

Golden stars had been engraved on the funeral steles of the firefighters: here lay Pravik, Kibenok, Ignatenko, Vashchuk, Tishchura, Titenok.

There were no marks of distinction, however, on the gravestones of the nuclear operators. None of the photographs that had originally been placed on the graves now remained: that of Leonid Toptunov was the sole exception. He was still only a boy in the picture, with round chubby cheeks and a small mustache. A handsome, well-kept bench, installed by his father, stood next to the grave. It seemed to me that Toptunov's grave was the most lovingly maintained.

Twenty-six graves, six of them the last resting place of the heroic firefighters. The remaining twenty contained the operators of No. 4 unit, electrical engineers, turbine engineers, and adjusters. Two women—Klavdia Ivanovna Luzganova and Yekaterina Aleksandrovna Ivanenko—had been assigned to a paramilitary security service at the plant. One of them, on duty in a passage opposite No. 4 unit, stayed at her post all night. The second was in the spent fuel depository which was still under construction, over 300 yards from the unit. Those graves also contained real heroes whose bravery saved the plant quite as much as that of the firefighters. I have mentioned them before. They are Vershinin, Novik, Brazhnik, and Perchuk, machinists from the turbine hall, who extinguished the interior fires which, had they spread, would have devastated the entire nuclear power station. How were they rewarded? To my knowledge, they have not been nominated for any decorations at all. Neither has Valery Ivanovich Perevozchenko, who moved heaven and earth to save those under his command and to rescue them from the high radiation zone.

Another person not nominated for a decoration is Anatoly Andreyevich Sitnikov, who exposed himself to mortal danger in order to find out exactly what had happened to No. 4 reactor.

Although he could have left and thus survived, Georgi Illarionovich Popov, an adjuster specializing in vibrations, who had come from Kharkov and just happened to be present that night, did not leave the turbine hall and did his utmost to help the turbine staff put out the fire there. Yet he has not been nominated for a decoration.

No decoration is planned for Anatoly Ivanovich Baranov, an electrical engineer, who together with Lelechenko confined the emergency in the electrical equipment by disconnecting the supply of hydrogen to the generator and restored current to No. 4 unit in the midst of extreme gamma radiation fields.

Lelechenko is buried in Kiev. He was awarded the Order of Lenin posthumously.

Another point needs to be made in connection with decorations. The documents dealing with decorations for nuclear operators were processed in the most dreadful secrecy. Why? I, for one, see absolutely no reason for it, especially as some genuine heroes, who should inspire pride in the living, were not decorated at all. Their families, children, and grandchildren must be really proud of them.

I believe that justice will triumph. Heroism cannot be hidden.

I walked past the graves, stopping for some time at each. I laid flowers on the gravestones. The firefighters and six nuclear operators died agonizing deaths between 11 May and 17 May 1986. They received the highest doses of radiation and absorbed the greatest quantities of radionuclides. Their bodies were radioactive and, as we have seen, were buried, on the insistence of the epidemiological center, in soldered lead coffins. I found that rather sad, as it prevented the earth from performing its eternal and necessary function—that of turning the bodies of the dead into dust. Such is the power of the atom! Even death and burial are not the same as for ordinary people. Ancient funerary traditions are thereby broken, and a human burial is rendered impossible.

Nonetheless, I say to them: rest in peace, sleep soundly. Your deaths jolted everybody out of their complacency. As a result, if only briefly, people shifted away from their unquestioning and routine compliance with orders.

A NEW CULTURE
FOR THE NUCLEAR AGE

But so much still remains to be done! What further lessons still need to be learned! What a battle must be fought in order to make our earth truly clean and safe for life and happiness!

Meanwhile, the nuclear bureaucrats are not asleep. Though somewhat bruised by the Chernobyl explosion, they are once again rearing their heads, praising the completely "safe" power of the peaceful atom, while not forgetting to cover up the truth. For it is not possible to sing the praises of the peaceful atom unless the truth is covered up. By this I mean the truth about the complexity and dangers of the work done by people in the nuclear power industry, and the potential hazards of nuclear power stations for the environment and for the general population, which knows nothing about radiation.

On 18 July 1986, Mayorets, minister of energy and electrification of the USSR, issued instructions strictly forbidding his subordinates to tell the truth about Chernobyl in the press or on radio or television. What, one wonders, was the minister afraid of? It is obvious. He was afraid of losing his job. But did he really need to worry? He will leave it, as he acquired it, by an act of his own will. Lacking both knowledge and experience, he had no right to be in such a job in the first place.

But he will not give up so easily. There is no point in hoping. Though the sooner he does quit, the better it will be. All of us need the truth, the whole truth, and nothing but the truth, for reasons that I hope to have made clear.

Here I would like to quote some eminently sensible passages from an article written by an American nuclear scientist, Karl Z. Morgan* who sounds a note of caution. I would willingly quote similar passages by academicians A. P. Aleksandrov or Ye. P. Velikhov, for example, but they have not written any.

Here is what Karl Z. Morgan has to say:

At present there is a preponderance of evidence that there is no threshold dose of ionizing radiation so low that it is safe or such that the risk of damage (even serious damage such as leukemia) is zero. . . .

Releases of radioactive noble gas is the principal source of population exposure from routine operation of BWR's boiling water reactors.†

The author then singled out krypton-85, which has a half-life of 10.7 years, as being particularly dangerous. He went on:

I would like to register a strong complaint regarding a practice that has developed in the nuclear energy industry, of "burning" and even "burning out" temporary employees. By this we mean employing poorly instructed and untrained persons temporarily to carry out "hot" jobs. Because of . . . a lack of appreciation of the risks of chronic exposure, such employees are much more likely to be involved in radiation accidents that could result in harm to themselves and others. I consider the practice of burning out employees to be highly immoral and unless the nuclear energy industry desists from such practices, I (and I am afraid many others) will cease to be strong supporters of this industry. . . .

During the past 10 to 15 years new data have indicated the risk of radiation induced cancer in humans is 10 or more times what we considered it to be in 1960 and there is no evidence of the existence of a safe threshold dose.

I should like to finish this chronicle by quoting a distinguished Soviet scientist, a full member of the Soviet Academy of Medical

*Karl Z. Morgan, "Ways of Reducing Radiation Exposure in a Future Nuclear Power Economy," from *Nuclear Power Safety*, edited by James H. Rust and Lynn E. Weaver (New York, Pergamon Press, 1976), pp. 156, 160-162. Russian edition published by Atomizdat, Moscow, 1980.
†A noble gas is inert: it neither reacts chemically nor is absorbed into body tissues, although it can enter the body by inhalation through the lungs.—Ed.

Science and a major specialist on leukemia, Andrei Ivanovich Vorobyov. Here are his comments on the Chernobyl tragedy:

> You can imagine what would happen to the planet if nuclear power stations were bombed, even with conventional warheads, with no nuclear charges. The mere thought of mankind existing in such a mutilated state is surely intolerable to any civilized person. It seems to me that after this accident mankind must abandon its medieval mentality.
>
> Today there are many things that need to be re-evaluated. And although the number of fatalities was limited, and most of the injured will survive and recover, nonetheless the tragedy at Chernobyl has shown us just how bad a potential disaster could be. The thinking of all members of society must be entirely recast in a new mold—workers, scientists, whoever they may be. After all, accidents are never accidental. Everyone must now understand that life in the nuclear age demands the same kind of painstaking attention to detail as one finds in the calculation of a missile trajectory. The nuclear age cannot be nuclear in one area only, and nonnuclear everywhere else. It is vitally important to realize that everyone must know, for example, what a chromosome is. They must know this just as they know what a four-cylinder internal combustion engine is. It is impossible to live without this knowledge. Anyone who wants to live in the nuclear era has got to create a new culture, a whole new mindset.

I venture to hope that this book will help with the formation of that culture.

24 May 1987

Index

269